Analysis of Electrical Machines

Authored by Valéria Hrabovcová, Pavol Rafajdus and Pavol Makyš

Published in London, United Kingdom

IntechOpen

Supporting open minds since 2005

Analysis of Electrical Machines
http://dx.doi.org/10.5772/intechopen.90135
Authored by Valéria Hrabovcová, Pavol Rafajdus and Pavol Makyš

Contributors
Valeria Hrabovcova, Pavol Makyš, Pavol Rafajdus

First published in London, United Kingdom, 2020 by IntechOpen
IntechOpen is the global imprint of INTECHOPEN LIMITED, registered in England and Wales, registration number: 11086078, 7th floor, 10 Lower Thames Street, London, EC3R 6AF, United Kingdom
Printed in Croatia

British Library Cataloguing-in-Publication Data
A catalogue record for this book is available from the British Library

Additional hard and PDF copies can be obtained from orders@intechopen.com

Analysis of Electrical Machines
Authored by Valéria Hrabovcová, Pavol Rafajdus and Pavol Makyš
p. cm.
Print ISBN 978-1-83880-207-3
Online ISBN 978-1-83880-208-0
eBook (PDF) ISBN 978-1-83962-687-6

An electronic version of this book is freely available, thanks to the support of libraries working with Knowledge Unlatched. KU is a collaborative initiative designed to make high quality books Open Access for the public good. More information about the initiative and links to the Open Access version can be found at www.knowledgeunlatched.org

For EU product safety concerns:
IN TECH d.o.o., Prolaz Marije Krucifikse Kozulić 3, 51000 Rijeka, Croatia, info@intechopen.com or visit our website at intechopen.com.

Meet the authors

Valéria Hrabovcová is Professor Emeritus of Electrical Machines at the University of Žilina, Faculty of Electrical Engineering and Information Technology, Slovakia. Her professional and research interests cover all kinds of electrical machines, including electronically commutated electrical machines. She has worked on many research and development projects and has written numerous scientific publications and monographs in the field of electrical engineering. Her work also includes various pedagogical activities, and she has participated in many international educational projects.

Pavol Rafajdus is a professor of electrical machines at the University of Žilina, Faculty of Electrical Engineering and Information Technology, Slovakia. His research is focused on electrical machines, mainly switched reluctance motors and other electrical machine properties. He is the author or coauthor of more than 150 scientific papers focused on the design and optimization of the parameter analysis of electrical machines. He is supervisor of several PhD theses, which are mainly focused on electrical machines.

Pavol Makyš is an associate professor of electrical drives at the University of Žilina, Faculty of Electrical Engineering and Information Technology, Slovakia. His area of interest is electrical drives, focusing on unconventional control methods such as sensorless control, position control with high accuracy, adaptive control methods, and others. He has been a member of different research teams solving scientific and research tasks and projects led by universities and different industrial partners. He has been the supervisor of several bachelor, diploma, and PhD theses mainly focused on electrical drives applications.

Acknowledgements

Authors would like to thank for support by the R&D Operational Program, Centre of Excellence of Power Electronics Systems and Materials for their components II., No. OPVaV-2009/2.1/02-SORO and ITMS 26220120046 funded by the European Regional Development Fund (ERDF).

Contents

Preface

This university textbook is devoted to the students, doctorands, and post-doctorands of electrical engineering, to researchers and scientists dealing with design, optimization, and analysis of electrical machine properties. The authors used results of their pedagogical, science, and research activities in this area and hope that this book will be a useful instrument for all who have an interest to gain a deeper knowledge and wider overview in the area of electrical machine analysis.

The purpose of this book is to present methods that are used for analysis of electrical machine properties in transients and steady-state conditions, by means of simulation in MATLAB-Simulink system and finite element method (FEM). To be able to simulate transients and steady-state condition, the parameters of the electrical machine equivalent circuit must be known.

Therefore, in the three main chapters, the next methods are presented: (1) a method in which the parameters (resistances and inductances) are calculated on the basis of geometrical dimensions and material properties as it is usually made in the design process; (2) a method of the general theory of electrical machines, by means of which the transients are investigated in two perpendicular axes; and (3) FEM, which as a mathematical method in application to electrical machines is an excellent instrument to investigate many of their properties. It enables to investigate properties of the manufactured machines in great detail, if their geometrical dimensions and material properties are known. All these methods are suitable also for a prefabricated period during which it is possible to optimize their properties. It is possible to measure parameters and characteristics in transient and steady-state conditions, after the machines are fabricated, and in this way to confirm correctness of the design calculation.

All three methods are explained and described in great detail, including simulation application in the MATLAB-Simulink system for each kind of the known machines, with suitable examples of simulations and time waveforms. In the annexes, simulation block diagrams devoted to the beginners, to be able to carry out their first own implementations of the simulation models, are shown.

The authors thank the reviewers for the careful reading of the manuscript and precious advices and recommendations that helped increasing the quality of this book.

Valéria Hrabovcová, Pavol Rafajdus, Pavol Makyš
University of Žilina,
Žilina,
Slovakia

Analysis of Electrical Machines

Valéria Hrabovcová, Pavol Rafajdus and Pavol Makyš

Abstract

The aim of this book is to present methods needed for the analysis of electrical machine performances in transients and steady-state conditions by means of simulations in MATLAB-Simulink and by means of FEM. The parameter determination based on the design procedure is presented, and some examples are given. The authors used outputs of their long-time experiences in research and scientists work as university professors and use their pedagogical skills to create a guide on how to investigate parameters of equivalent circuits and apply them for investigation of transients and steady-state conditions.

The book is devoted to the postgraduate and PhD students, post PhD students, researchers, and scientist workers in electrical engineering, which deal with design and analysis of electrical machines' performances and its optimization. The readers should be educated in theory and basic principles of operation of all electrical machines in steady-state conditions to be able to follow application of mathematical apparatus needed for transient investigation in general theory of electrical machines and FEM analysis.

The goal is to provide a guide for those who would like to learn how to investigate parameters of electrical machines and use them to design electrical machines and their optimization.

Keywords: general theory of rotating electrical machines, parameters of electrical machines and their calculation during their design, resistances, inductances, skin effect influence on the resistance and inductance, transients of electrical machines, parameters of equivalent circuits, FEM used to investigate parameters of electrical machines, induced voltage, electromagnetic torque, torque ripple, cogging torque

IntechOpen

Introduction

In a coincidence with the title of this book, we can start with explaining what a term of analysis means: in general, it is a system of methods by means of which properties of investigated matters are gained. Here the properties of electrical machines are analyzed; therefore, it is welcomed to introduce the methods on how to do it.

To proceed in the investigation of transients and steady-state condition, it is necessary to know equivalent circuit parameters (resistances and inductances). The first method that is given below in Chapter 1 is the method of parameters calculated based on the design process in which geometrical dimensions and material properties must be known (see [1–3]). The other method on how to get the equivalent circuit parameters is to make measurements and testing, but this can be done only on the fabricated pieces. It is a very welcomed method on how to verify the calculated parameters gained during the design process. The method of measurement is not given here and can be found, e.g., in [4–6].

The other approach is if there is a real fabricated machine but without any documents and data. Then it is very useful to make the so-called inverse design calculation. It means to take all geometrical dimensions which can be seen on the real machine and get data from the real machine nameplate and catalogues, e.g., voltage, current, power, speed, pole numbers, slot shape, number of slots, number of conductors in the slot, etc. and to continue in the calculation of air gap magnetic flux density, etc. to the required parameters.

The general theory of electrical machines is presented in Chapter 2. It is possible to investigate transients and steady-state conditions of electrical machines by means of this theory, see [7–9]. The transients are solved on the basis of differential equations in which the parameters of equivalent circuits, i.e., resistances and inductances, are needed. The accuracy and reliability of the simulation results depend on the accuracy of the parameter values. Therefore, determination of the parameters must be done with the highest care.

Chapter 3 is formulated in this sense. This chapter is devoted to the modern computer method called finite element method (FEM), see also [10]. This method enables to investigate not only some parameters, mainly magnetizing inductances, but also the other properties such as losses, air gap developed torque, ripple torque, efficiency, etc., see [11–13].

It is important to add that verification of the calculated and simulated values and waveforms is made by measurement on a real machine if it exists. It is recommended to create a reliable simulation model. It means the gained simulated outputs verify on a real machine, and if the coincidence of measured and simulated values is satisfactory, such model can be employed to optimize geometrical dimensions or a concrete configuration, like slot shape, rotor barriers, etc., and the simulation outputs are considered as reliable. In such a way, it is possible in pre-manufacturing period to optimize the construction of the machine to the required properties, e.g., maximal torque, minimal ripple torque, maximum efficiency, and so on.

The authors would like to point out that all simulation models of electrical machines analyzed in this book are shown at the end as Appendixes A, B, C, and D, in the MATLAB-Simulink program. These models were employed in the appropriate chapters during the investigation of the simulation waveforms of all electrical machines.

Chapter 1

Parameters of the Rotating Electrical Machines

1.1 Inductances

In the rotating electrical machines, the total magnetic flux can be divided into two components: main flux (air gap flux) and leakage flux. The main flux enables electromagnetic energy conversion, but a proportion of the total flux does not participate in energy conversion, and this part is called leakage flux. The main flux must cross the air gap of rotating machine and its function is electromagnetically connected to both stator and rotor windings. The leakage flux is linked only with this winding in which it was created.

The main magnetic flux in air gap Φ_{μ} creates linkage flux ψ_{μ} and corresponds to the magnetizing inductance L_{μ}. The leakage flux Φ_{σ} creates leakage linkage flux ψ_{σ} and corresponds to the leakage inductance L_{σ}. In the case of induction machines, stator inductance is the sum of magnetizing and stator leakage inductance: $L_{s} = L_{\mu} + L_{\sigma s}$. In the case of synchronous machines, this inductance is called synchronous inductance. In the next part, all components of the inductances are investigated.

1.1.1 Magnetizing inductance

Magnetizing inductance is the most important inductance of electrical machine and is defined by the winding, geometrical dimensions of the magnetic circuit, and the employed materials. It is derived from m-phase machine on the basis of magnetic flux density distribution on the surface of the rotor and its maximal value $B_{\delta\max}$ (**Figure 1**), on the span of the pole pitch τ_{p}, and the reduced length of the machine $l' = l_{Fe} + 2\delta$.

This relationship is valid for machines without ventilating ducts and takes into account the magnetic flux distribution on the end of the machines [1].

Maximal value of the air gap magnetic flux can be calculated as a surface integral of magnetic flux density B on one pole surface S:

$$\Phi_{\max} = \int_{S} B dS = \alpha_{i} \tau_{p} l' B_{\delta\max}, \tag{1}$$

where α_{i} is a ratio of the arithmetic average value of the magnetic flux density B_{av} and maximal value of the magnetic flux density:

$$\alpha_{i} = \frac{B_{av}}{B_{\delta\max}}. \tag{2}$$

The sinusoidal distributed waveform of the magnetic flux density is $\alpha_{i} = 2/\pi$.

An expression "maximal value of the magnetic flux" means maximal flux which penetrates the surface created by the coil and therefore creates one phase maximal linkage flux of the winding on the stator (the subscript "ph" is used), with number of turns N_{s} and winding factor k_{ws}:

$$\psi_{\mu ph} = N_{s} k_{ws} \Phi_{\mu} = N_{s} k_{ws} \alpha_{i} \tau_{p} l' B_{\delta\max}. \tag{3}$$

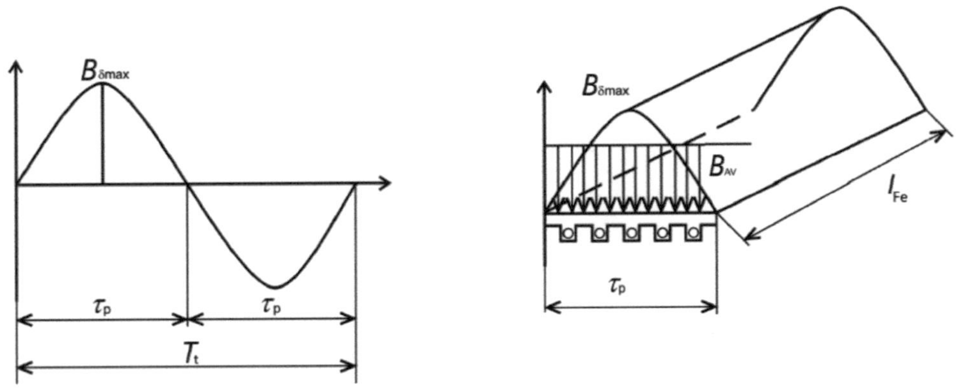

Figure 1.
Distribution of the magnetic flux density fundamental harmonic in the air gap over a pole pitch and on the length of the l_{Fe}.

Magnitude of air gap magnetic flux density can be expressed on the basis of current linkage and the relationship between magnetic flux density and intensity of magnetic field. In the most simple case, $B = \mu H, Hl = NI, H = NI/l$, where the permeability of vacuum is taken and the length of the magnetic force line is only the length of air gap. The source of magnetomotive force (current linkage) is expressed for fundamental harmonic of rectangular waveform of the single-phase winding:

$$U_{\text{magmaxph}} = \frac{4}{\pi}\frac{N_s k_{ws}}{2p}\sqrt{2}I_s. \tag{4}$$

Then the magnitude of the magnetic flux density is:

$$B_{\delta\text{max}} = \frac{\mu_0}{\delta_{ef}} U_{\text{magmaxph}} = \frac{\mu_0}{\delta_{ef}}\frac{4}{\pi}\frac{N_s k_{ws}}{2p}\sqrt{2}I_s, \tag{5}$$

and after the substitution to the expression of flux linkage, it yields:

$$\psi_{\mu\text{ph}} = N_s k_{ws}\alpha_i \tau_p l' \frac{\mu_0}{\delta_{ef}}\frac{4}{\pi}\frac{N_s k_{ws}}{2p}\sqrt{2}I_s = \alpha_i \tau_p l' \frac{\mu_0}{\delta_{ef}}\frac{4}{\pi}\frac{(N_s k_{ws})^2}{2p}\sqrt{2}I_s. \tag{6}$$

By dividing the result by the peak value of the current, which in this case is magnetizing current, we obtain the magnetizing inductance of a single-phase winding (the main inductance):

$$L_{\mu\text{ph}} = \alpha_i \frac{\mu_0}{\delta_{ef}}\frac{4}{\pi}\frac{1}{2p}\tau_p l'(N_s k_{ws})^2 = \alpha_i \frac{\mu_0}{\delta_{ef}}\frac{2}{\pi}\frac{(N_s k_{ws})^2}{p}\tau_p l'. \tag{7}$$

If the winding is multiphase, the magnetic flux is created by all the phases of the winding, with the corresponding instantaneous values of the currents. As it is known, the three-phase winding creates the magnitude of the air gap magnetic flux density equal to 1.5 multiple of that which is created by the single phase. The magnetizing inductance of an m-phase winding can be calculated by multiplying the main inductance by $m/2$:

$$L_\mu = \frac{m}{2}\alpha_i \frac{\mu_0}{\delta_{ef}}\frac{2}{\pi}\frac{(N_s k_{ws})^2}{p}\tau_p l' = \alpha_i m \frac{\mu_0}{\delta_{ef}}\frac{\tau_p l'}{\pi}\frac{(N_s k_{ws})^2}{p}, \tag{8}$$

and after the substitution of pole pitch, the expression is obtained, in which its dependence on the parameters and geometrical dimensions of the machine is seen:

$$L_\mu = \alpha_i m \frac{\mu_0}{\delta_{ef}} \frac{D_\delta}{\pi} \frac{(N_s k_{ws})^2}{p^2} l'. \quad (9)$$

Here it is seen that the magnetizing inductance depends on magnetic circuit saturation, i.e., α_i, effective air gap δ_{ef}, in which the Carter factor and saturation of magnetic circuit is included, on the length of the machine l', phase number m, and quadrate of the effective number of the turns $N_s k_{ws}$. In Eq. (9), it is seen that the magnetizing inductance is inversely proportional to p^2, which means that in the case of asynchronous motors where it is important to get L_μ as high as possible, it is not suitable to employ multipole arrangement. In the case of the synchronous machines, the developed torque is inversely proportional to the synchronous inductance and also to the magnetizing inductance. Therefore, in the synchronous machines, there are multipole arrangements with lower inductance ordinary.

Effective air gap δ_{ef} includes Carter factor as well as the effect of the magnetic circuit saturation. This influence is in the interval of some to 10%. In such a case the magnetic circuit is already considerably saturated. In very precisely designed induction, motors can be the current linkage (magnetomotive force) needed for iron parts of magnetic circuit greater than for the air gap. On the other side in synchronous machines with permanent magnets, in which the equivalent air gap includes in d-axis also the length of permanent magnets, the value of current linkage needed for iron parts is very small.

Magnetizing inductance is not constant but depends on the voltage and the torque. The higher voltage activates the higher magnetic flux density; this activates higher saturation of magnetic circuit, and this requires higher magnetizing current.

1.1.2 Leakage inductance

Leakage inductances are described by the leakage magnetic fields, which are linked only with the turns of that winding by which they were created. It means they do not cross the air gap.

In greater detail, it can be said that leakage magnetic fluxes include the following:

- All components of magnetic field that do not cross air gap
- Those components of magnetic field that cross the air gap, but they do not take part in the electromechanical energy conversion

The leakage fluxes that do not cross the air gap can be divided into the next components:

a. Slot leakage flux $\approx L_{\sigma d}$ (slot leakage inductance)

b. Tooth tip leakage flux $\approx L_{\sigma z}$ (leakage tooth tip inductance)

c. End winding leakage flux $\approx L_{\sigma ew}$ (end winding leakage inductance)

d. Pole leakage flux $\approx L_{\sigma p}$ (pole leakage inductance)

Leakage fluxes that do not cross the air gap are included into the air gap magnetic flux $\Phi_\delta \approx L_\delta$ (air gap leakage inductance). Air gap magnetic flux do not link completely windings of the machine because of short pitching, slot skewing, and spatial distribution of the winding, causing air gap harmonic components in the air gap, and do not contribute to the electromechanical energy conversion. The weaker linking between the stator and rotor windings caused by the short pitching and slot skewing is taken into account by means of the pitch winding factor k_p and skewing factor k_{sq}.

According to the electrical motor design tradition, leakage inductance L_σ can be divided into the following partial leakage inductance: skew leakage inductance $L_{\sigma sq}$, air gap leakage inductance, slot leakage inductance, tooth tip leakage inductance, and end winding leakage inductance. The leakage inductance of the machine is the sum of these leakage inductances:

$$L_\sigma = L_{\sigma sq} + L_{\sigma\delta} + L_{\sigma d} + L_{\sigma z} + L_{\sigma ew}. \tag{10}$$

1.1.2.1 Skew leakage inductances

Skewing slot factor defines skewing leakage inductance:

$$L_{\sigma sq} = \sigma_{sq} L_\mu, \tag{11}$$

where the factor of the skewing leakage σ_{sq} is given by the skewing factor k_{sq}:

$$\sigma_{sq} = \frac{1 - k_{sq}^2}{k_{sq}^2}.$$

At the skewing by one slot, this factor is given by the expression:

$$k_{sq} = \frac{sin\,\frac{\pi}{2}\frac{1}{mq}}{\frac{\pi}{2}\frac{1}{mq}}, \tag{12}$$

where q is the number of slots per phase per pole, and it is possible to calculate for each υ—harmonic component:

$$k_{sq\upsilon} = \frac{sin\,\upsilon\frac{\pi}{2}\frac{1}{mq}}{\upsilon\frac{\pi}{2}\frac{1}{mq}}. \tag{13}$$

Example 1. Calculate skewing factors for the fundamental stator slot harmonic component in four-pole rotor cage induction motor with 36 slots on the stator when the rotor slots are skewed by one stator slot.

Solution: As it is known, m-phase winding creates harmonic components of the order $\upsilon = 1 \pm 2\,cm$, where c = 0, 1, 2, 3, and so on.

The number of stator slots per phase per pole $q = 36/(3\cdot 4) = 3$. The first stator slot harmonics are $(1 \pm 2\,mqc) = 1 \pm 2\cdot 3\cdot 3\cdot c = -17, 19, -35, 37,$ if c = 1, 2. The skewing factor according to Eq. (13) for the fundamental and further harmonics is:

υ	1	−17	19	−35	37
k_{sq}	0.995	0.06	−0.05	−0.03	0.03

It can be seen that the lowest order stator harmonics (−17, 19, −35, 37) have very small skewing factors, and thereby their effects are eliminated to a great degree (3–6%). The fundamental harmonic is reduced only by 0.5%.

1.1.2.2 Air gap leakage inductance

Electromotive force (induced voltage) is given by the magnetizing inductance as a result of a propagating fundamental component of air gap flux density. Because of a spatial slotting and winding distribution, the permeance harmonics induce voltage of fundamental frequency of the winding. The air gap leakage inductance, i.e., the harmonic leakage inductance components, takes this into account. In integer slot machines per phase per pole q, the air gap leakage remains usually low. But in the case of fractional slot machines mainly in the machines with concentrating coils wound around the tooth, its influence can be even dominating. It relates to machines with permanent magnets. This case can be studied in greater details in [1].

The sum of all induced voltages from all harmonic components, fundamental included, gives the basis for the calculation of the total inductance, which is the sum of magnetizing and air gap leakage inductance:

$$L_{\mu} + L_{\sigma\delta} = \frac{\mu_0}{\pi}\frac{m}{\delta}Dl'\left(\frac{N}{p}\right)^2 \sum_{\upsilon=-\infty}^{\upsilon=+\infty}\left(\frac{k_{w\upsilon}}{\upsilon}\right)^2. \tag{14}$$

The expression for the fundamental harmonic, i.e., $\upsilon = 1$, represents fundamental component; it means magnetizing inductance of the machine L_{μ}, which is calculated on the basis of Eq. (9), if $\alpha_i = 2/\pi$ is introduced. The rest of the equation represents air gap inductance ($\upsilon = 1$ is omitted):

$$L_{\sigma\delta} = \frac{\mu_0}{\pi}\frac{m}{\delta}Dl'\left(\frac{N}{p}\right)^2 \sum_{\substack{\upsilon=-\infty \\ \upsilon\neq 1}}^{\upsilon=+\infty}\left(\frac{k_{w\upsilon}}{\upsilon}\right)^2. \tag{15}$$

The air gap leakage inductance can be expressed also by means of the air gap leakage factor:

$$\sigma_{\delta} = \sum_{\substack{\upsilon=-\infty \\ \upsilon\neq 1}}^{\upsilon=+\infty}\left(\frac{k_{w\upsilon}}{\upsilon k_{w1}}\right)^2. \tag{16}$$

Then the leakage inductance is given by:

$$L_{\sigma\delta} = \sigma_{\delta}L_{\mu}. \tag{17}$$

Of course, in Eq. (16), only the harmonics that are created by the given winding are used.

1.1.2.3 Slot leakage inductance

This inductance is created by a real leakage flux, which is closed through the space of the slot. Magnetic permeance of the magnetic circuit is taken as infinite; therefore the length of the magnetic circuit force line is taken only by the width of the slot (in the slot there is nonmagnetic material, i.e., there the vacuum permeability is used). For the rectangular slot, the magnetic permeance is derived in [1]. Gradually integrating the magnetic force lines and magnetic flux density along the slot height h in the area where the current flows (**Figure 2**), the magnetic permeance of the slot for the slot leakage is obtained:

$$\Lambda_{\text{mag}} = \mu_0 l' \frac{h_4}{3b_4}. \tag{18}$$

The permeance factor λ is defined, because in each slot the vacuum permeability and the length of the machine act:

$$\lambda = \frac{\Lambda_{\text{mag}}}{\mu_0 l'}. \tag{19}$$

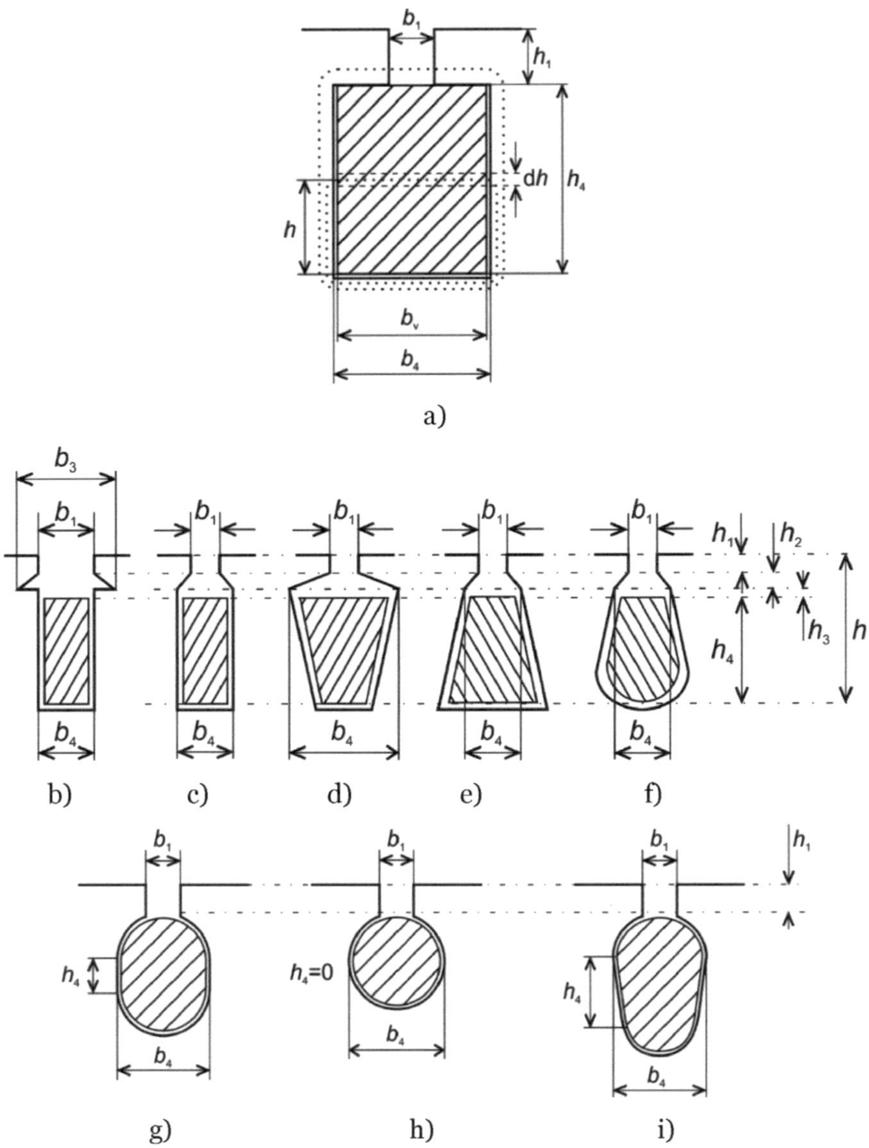

Figure 2.
Geometrical dimensions of various slot shapes to define their permeance factors, the calculations of which are given in the text.

For the rectangular slot (**Figure 2a**) with the slot width b_4 and the slot height h_4, the permeance factor is:

$$\lambda_4 = \frac{h_4}{3b_4}. \tag{20}$$

In the slot area with the height h_1, there is no current; therefore:

$$\lambda_1 = \frac{h_1}{b_1} \tag{21}$$

The sum $\lambda_1 + \lambda_4 = \lambda_\mathrm{d}$ is the permeance factor of the whole slot, and the leakage inductance of the slot is:

$$L_{\sigma\mathrm{d}} = \frac{4m}{Q}\mu_0 l' N^2 \lambda_\mathrm{d} = 2\frac{N^2}{pq}\mu_0 l' \lambda_\mathrm{d}, \tag{22}$$

where Q is the number of slots around the machine periphery, $2p$ is the number of poles, q is the number of slots per pole per phase, and N is the phase number of turns.

Equation (22) is derived from [1] and other references dealing with the machine design.

The expressions gained on the basis of magnetic permeance integration along the slot height of single-layer winding according to **Figure 2** are given.

For the slot shapes b, c, d, e, and f, the permeance factor will be calculated as follows:

$$\lambda_\mathrm{d} = \frac{h_4}{3b_4} + \frac{h_3}{b_4} + \frac{h_1}{b_1} + \frac{h_2}{b_4 - b_1} \, ln \, \frac{b_4}{b_1}. \tag{23}$$

For the slot in **Figure 2g**, the next expression is known:

$$\lambda_\mathrm{d} = \frac{h_4}{3b_4} + \frac{h_1}{b_1} + 0.685, \tag{24}$$

and for the round slot from **Figure 2h**, the next expression is valid:

$$\lambda_d = 0.47 + 0.066\frac{b_4}{b_1} + \frac{h_1}{b_1}. \tag{25}$$

The slot leakage inductance of the double-layer winding, according to [1], on the basis of **Figure 3**, the appropriate expression is derived. It is necessary to consider that in some slots there are coil sides of different phases.

$$\lambda_\mathrm{d} = k_1\frac{h_4 - h'}{3b_4} + k_2\frac{h_1}{b_4} + \frac{h'}{4b_4}, \tag{26}$$

where

$$k_1 = \frac{5 + 3g}{8}, \tag{27}$$

$$k_2 = \frac{1 + g}{2}. \tag{28}$$

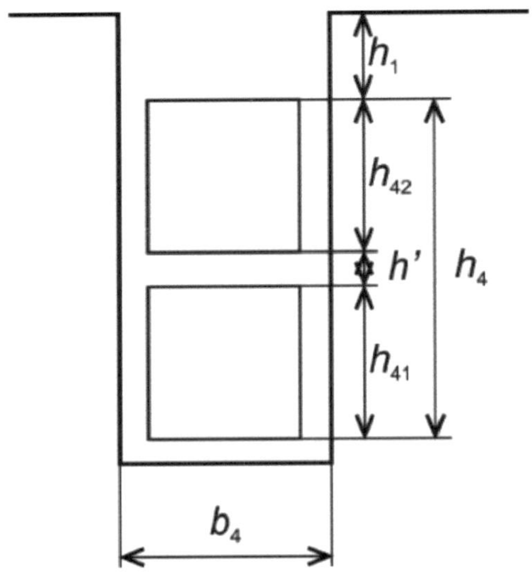

Figure 3.
Geometry of the slot with double-layer winding.

The factor g is linked with the fact that in double-layer winding with short pitching, the coil sides of upper and lower layers in some slots belong to different phases. If the phase shifting of the upper and lower layers is γ, the total current linkage must be multiplied by $\cos\gamma$. Because the phase shifting can be different in each slot, the average value g on the $2q$ coil sides is:

$$g = \frac{1}{2q}\sum_{n=1}^{2q} \cos\gamma_n,$$

but the factors k_1 and k_2 can be calculated also on the basis of short pitching (see below).

Similar to Eq. (26), also the equation for double-layer winding can be written for the slots in **Figure 2** from (b) till (f):

$$\lambda_{\mathrm{d}} = k_1 \frac{h_4 - h'}{3b_4} + k_2\left(\frac{h_3}{b_4} + \frac{h_1}{b_1} + \frac{h_2}{b_4 - b_1} ln \frac{b_4}{b_1}\right) + \frac{h'}{4b_4}, \tag{29}$$

and for slots from (g) till (i):

$$\lambda_{\mathrm{d}} = k_1 \frac{h_4 - h'}{3b_4} + k_2\left(\frac{h_1}{b_1} + 0.66\right) + \frac{h'}{4b_4}. \tag{30}$$

These expressions are valid also for the slots with single-layer winding, if $h' = 0$ and $k_1 = k_2 = 1$ are introduced.

If the winding is short pitching , factors k_1 and k_2 can be calculated by means of short pitching Y/Q_{p}, where Y is a real pitch and Q_{p} is a full pitch (the pole pitch is expressed by means of number of slots per pole), in this form:

For three-phase winding:

$$k_1 = 1 - \frac{9}{16}\varepsilon, k_2 = 1 - \frac{3}{4}\varepsilon, \tag{31}$$

where

$$\varepsilon = 1 - \frac{Y}{Q_p}. \tag{32}$$

For two-phase winding:

$$k_1 = 1 - \frac{3}{4}\varepsilon, k_2 = 1 - \varepsilon. \tag{33}$$

Example 2. Calculate slot leakage inductance of a double-layer winding if $2p = 4$, $m = 3$, $Q = 24$, $Y/Q_p = 5/6$, and $N = 40$. The slot shape and dimensions are according to **Figures 2c** and **3** as follows:

$b_1 = 0.003$ m, $h_1 = 0.002$ m, $h_2 = 0.001$ m, $h_3 = 0.001$ m, $h' = 0.001$ m, $b_4 = 0.008$ m and $h_{41} = h_{42} = 0.009$ m, ($h_4 = 0.019$ m), $l' = 0.25$ m.

Compare the results with the slot leakage inductance of a corresponding double-layer full-pitch winding.

Solution: The short pitching is $\varepsilon = 1/6$, and $k_1 = 1 - \frac{9}{16}\frac{1}{6} = 0.906$, $k_2 = 1 - \frac{3}{4}\frac{1}{6} = 0.875$. The permeance factor is according to Eq. (29):

$$\begin{aligned}\lambda_d &= k_1\frac{h_4 - h'}{3b_4} + k_2\left(\frac{h_3}{b_4} + \frac{h_1}{b_1} + \frac{h_2}{b_4 - b_1}\ln\frac{b_4}{b_1}\right) + \frac{h'}{4b_4} \\ &= 0.906\frac{0.018}{3\cdot 0.008} + 0.875\left(\frac{0.001}{0.008} + \frac{0.002}{0.008} + \frac{0.001}{0.008 - 0.003}\ln\frac{0.008}{0.003}\right) \\ &\quad + \frac{0.001}{4\cdot 0.008} \\ &= 1.211\end{aligned}$$

and slot leakage inductance is according to Eq. (22):

$$\begin{aligned}L_d &= \frac{4m}{Q}\mu_0 l' N^2 \lambda_d = \frac{4\cdot 3}{24}4\pi\cdot 10^{-7}\cdot 0.25\cdot 40^2\cdot 1.211 = 0.2513\cdot 10^{-3}\cdot 1.211 \\ &= 0.304\text{ mH},\end{aligned}$$

for a double-layer full-pitch winding $k_1 = k_2 = 1$ and Eq. (29) yields:

$$\begin{aligned}\lambda_d &= k_1\frac{h_4 - h'}{3b_4} + k_2\left(\frac{h_3}{b_4} + \frac{h_1}{b_1} + \frac{h_2}{b_4 - b_1}\ln\frac{b_4}{b_1}\right) + \frac{h'}{4b_4} \\ &= \frac{0.018}{3\cdot 0.008} + \left(\frac{0.001}{0.008} + \frac{0.002}{0.008} + \frac{0.001}{0.008 - 0.003}\ln\frac{0.008}{0.003}\right) + \frac{0.001}{4\cdot 0.008} \\ &= 1.352\end{aligned}$$

The slot leakage inductance is now:

$$L_{\sigma d} = \frac{4m}{Q}\mu_0 l' N^2 \lambda_d = 0.2513\cdot 10^{-3}\cdot 1.352 = 0.340\text{ mH}$$

It is seen that the phase shift of the different phase coil sides in the double-layer winding causes a smaller slot leakage inductance for the short-pitched winding than the full-pitch winding. The slot leakage inductance in this case is about 10% smaller for the short-pitched winding.

1.1.2.4 Tooth tip leakage inductance

The tooth tip leakage inductance is determined by the magnitude of leakage flux flowing in the air gap outside the slot opening. This flux is illustrated in **Figure 4**. The current linkage in the slot causes a potential difference between the teeth on opposite sides of the slot opening, and as a result a part of the current linkage will be used to produce the leakage flux of the tooth tip.

Tooth tip leakage inductance can be determined by applying a permeance factor:

$$\lambda_z = k_2 \frac{5\left(\frac{\delta}{b_1}\right)}{5 + 4\left(\frac{\delta}{b_1}\right)}, \tag{34}$$

where $k_2 = 1 - \frac{3}{4}\varepsilon$ is given by Eq. (31). The tooth tip leakage inductance of the whole phase winding is given by applying Eq. (22):

$$L_{\sigma z} = \frac{4m}{Q}\mu_0 l' N^2 \lambda_z. \tag{35}$$

In the machines with salient poles, the air gap is taken at the middle of the pole, where the air gap is smallest. If the air gap is selected to be infinite, a limit value of 1.25 is obtained, which is the highest value for permeance factor λ_z. If the air gap is small, as in the case of asynchronous machines, the influence of the tooth leakage inductance is insignificant. The above given equations are not valid for the main poles of DC machines. The calculation for synchronous machines with permanent magnets is in Example 3.

Example 3. Calculate the tooth tip leakage of the machine in Example 2. The machine is now equipped with rotor surface permanent magnets neodymium-iron-boron of 8 mm thickness. There is a 2 mm physical air gap. Compare the result with the value of inductance in Example 2.

Solution: As the permanent magnets represent, in practice, air with relative permeability μ_{rPM} = 1.05, we may assume that the total air gap in the calculation of the tooth tip leakage is:

$$\delta = 2 + \frac{8}{1.05} = 9.62 \text{ mm}.$$

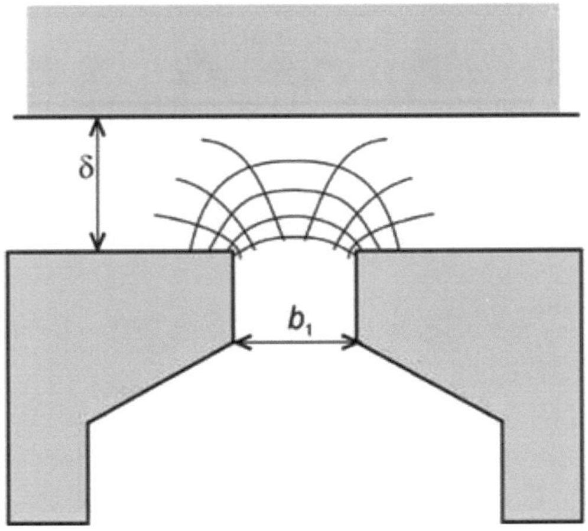

Figure 4.
Tooth tip flux leakage around a slot opening, creating a tooth tip leakage inductance.

The factor $k_2 = 1 - \frac{3}{4}\frac{1}{6} = 0.875$.
Then the factor of tip tooth permeance is:

$$\lambda_z = k_2 \frac{5\left(\frac{\delta}{b_1}\right)}{5 + 4\left(\frac{\delta}{b_1}\right)} = 0.875 \frac{5\left(\frac{0.00962}{0.003}\right)}{5 + 4\left(\frac{0.00962}{0.003}\right)} = 0.787,$$

and the tooth tip leakage inductance is:

$$L_{\sigma z} = \frac{4m}{Q}\mu_0 l' N^2 \lambda_z = \frac{4 \cdot 3}{24} 4\pi \cdot 10^{-7} \cdot 0.25 \cdot 0.787 \cdot 40^2 = 0.198 \text{ mH}.$$

In Example 2, the slot leakage inductance was 0.34 mH. As the air gap in a rotor surface magnet machine is long, the tooth tip leakage has a significant value, about 70% of the slot leakage inductance.

1.1.2.5 End winding leakage inductance

End winding leakage flux results from all the currents flowing in the end windings. The geometry of the end windings is usually difficult to analyze, and, further, all the phases of polyphase machines influence the occurrence of a leakage flux. Therefore, the accurate determination of an end winding leakage inductance would require three-dimensional numerical solution. On the other side, the end windings are relatively far from the iron parts; the end winding inductances are not very high. Therefore, it suffices to employ empirically determined permeance factor.

According to **Figure 5**, the end winding leakage flux is a result of influence of all coil turns belonging to the group coils q.

If according to Eq. (22) this q-multiple of the slot conductors is put and instead of the length of the machine, the length of the end winding l_{ew} is introduced, the equation for the end winding inductance calculation is as follows:

$$L_{\sigma ew} = \frac{2}{p}\mu_0 N^2 l_w \lambda_{ew}. \tag{36}$$

The average length of the end winding l_w and the product of $l_w\lambda_{ew}$ can be, according to **Figure 5**, written in the form:

$$l_w = 2l_{ew} + Y_{ew}, \tag{37}$$

$$l_w\lambda_{ew} = 2l_{ew}\lambda_{lew} + Y_{ew}\lambda_{Yew}, \tag{38}$$

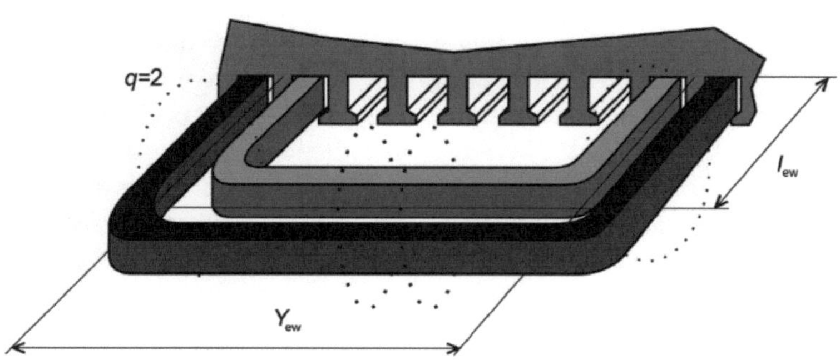

Figure 5.
Leakage flux and dimensions of the end winding.

where l_{ew} is the axial length of the end winding, measured from the iron laminations, and Y_{ew} is a coil span according to **Figure** 5. Corresponding permeance factors λ_{lew} and λ_{Yew} can be changed according to the type of stator and rotor windings and are shown in, e.g., [1] in Table 4.1 and 4.2. At a concrete calculation of the real machines, it can be proclaimed that Eq. (36) gives the sum of the leakage inductance of the stator and leakage inductance of the rotor referred to the stator and that the essential part 60–80% belongs to the stator.

Example 4. The air gap diameter of the machine in Example 2 is 130 mm, and the total height of the slots is 22 mm. Calculate end winding leakage inductance for a three-phase surface-mounted permanent magnet synchronous machine with $Q = 24$, $q = 2$, $N = 40$, $p = 2$, $l_{\mathrm{ew}} = 0.24$ m. The end windings are arranged in such a way that $\lambda_{\mathrm{lew}} = 0.324$ and $\lambda_{\mathrm{Yew}} = 0.243$.

Solution: Let us assume that the average diameter of the end winding is 130 + 22 = 152 mm. The perimeter of this diameter is about 480 mm. The pole pitch at this diameter is $\tau_{\mathrm{p}} = 480/4 = 120$ mm. From this it can be assumed that the width of the end winding is about the pole pitch subtracted by one slot pitch:

$$Y_{\mathrm{ew}} = \tau_{\mathrm{p}} - \tau_{\mathrm{d}} = 0.12 - \frac{0.48}{24} = 0.1\ \mathrm{m}.$$

and the length of the end winding is:

$$l_{\mathrm{ew}} = 0.5(l_{\mathrm{w}} - Y_{\mathrm{ew}}) = 0.5(0.24 - 0.1) = 0.07\ \mathrm{m}.$$

The product of the length and permeance factor is:

$$l_{\mathrm{w}}\lambda_{\mathrm{ew}} = 2l_{\mathrm{ew}}\lambda_{\mathrm{lew}} + Y_{\mathrm{ew}}\lambda_{\mathrm{Yew}} = 2 \cdot 0.07 \cdot 0.324 + 0.1 \cdot 0.243 = 0.07\ \mathrm{m}.$$

and end winding leakage inductance is:

$$L_{\sigma\mathrm{ew}} = \frac{2}{p}\mu_0 N^2 l_{\mathrm{w}}\lambda_{\mathrm{ew}} = \frac{2}{2}4\pi \cdot 10^{-7} \cdot 40^2 \cdot 0.07 = 0.1407\ \mathrm{mH}.$$

The slot leakage inductance of the 5/6 short-pitched winding in Example 2 is 0.304 mH, so it is seen that the slot leakage inductance is much higher than end winding leakage inductance.

1.2 Resistances

Not only inductances but also resistances are very important parameters of electrical machines. In many cases winding losses are dominant components of the total loss in electrical machines. The conductors in electrical machines are surrounded by ferromagnetic material, which at saturation can encourage flux components to travel through the windings. This can cause large skin effect problems, if the windings are not correctly designed. Therefore, this phenomenon must be considered.

It is convention to define at first the DC resistance R_{DC}, which depends on the conductivity of the conductor material σ_{c}, its total length l_{c}, cross-section area of the conductor S_{c}, and the number of parallel paths a in the winding without a commutator, per phase, or $2a$ in windings with a commutator:

$$R_{\mathrm{DC}} = \frac{l_{\mathrm{c}}}{\sigma_{\mathrm{c}} a S_{\mathrm{c}}}. \tag{39}$$

Resistance is highly dependent on the running temperature of the machine; therefore a designer should be well aware of the warming-up characteristics of the machine before defining the resistances. Usually the resistances are determined at the design temperature or at the highest allowable temperature for the selected winding type.

Windings are usually made of copper. The specific conductivity of pure copper at room temperature, which is taken 20°C, is σ_{Cu} = 58 × 10^6 S/m, and the conductivity of commercial copper wire is σ_{Cu} = 57 × 10^6 S/m. The temperature coefficient of resistivity for copper is α_{Cu} = 3.81 × 10^{-3}/K. Resistance at temperature t increased by Δt is $R_t = R(1 + \alpha_{Cu}\Delta t)$. The respective parameters for aluminum are σ_{Al} = 37 × 10^6 S/m, α_{Al} = 3.7 × 10^{-3}/K.

The accurate definition of the winding length in an electrical machine is a fairly difficult task. Salient-pole machines are a relatively simple case: the conductor length can be defined more easily when the shape of the pole body and the number of coil turns are known. Instead winding length of slot windings is difficult, especially if coils of different length are employed in the machine. Therefore, empirical expressions are used for the length calculation.

The average length of a coil turn of a slot winding l_{av} in low-voltage machines with round enameled wires is given approximately as:

$$l_{av} \approx 2l + 2.4Y_{ew} + 0.1 \text{ m}, \tag{40}$$

where l is the length of the stator stack and Y_{ew} is the average coil span (see **Figure 5**), both expressed in meters. For large machines with prefabricated windings, the following approximation is valid:

$$l_{av} \approx 2l + 2.8Y_{ew} + 0.4 \text{ m}. \tag{41}$$

When the voltage is between 6 and 11 kV, the next can be used:

$$l_{av} \approx 2l + 2.9Y_{ew} + 0.3 \text{ m}. \tag{42}$$

After the average length is determined, the DC resistance may be calculated according to Eq. (39), by taking all the turns and parallel paths into account.

1.3 Influence of skin effect on winding resistance and inductance

1.3.1 Influence of skin effect on winding resistance

The alternating current in a conductor and currents in the neighboring conductors create an alternating flux in the conductor material, which causes skin and proximity effects. In the case of parallel conductors, also circulating currents between them appear. The circulating currents between parallel conductors can be avoided by correct geometrical arrangement of the windings. In this chapter, the skin and proximity effects will be dealt together and called the skin effect.

Skin effect causes displacement of the current density to the surface of the conductor. If the conductor is alone in the vacuum, the current density is displaced in all directions equally to the conductor surface. But in the conductors embedded in the slots of electrical machines, the current density is displaced only in the direction to the air gap. In this manner, the active cross-section area of the conductors is reduced, increasing the resistance. This resistance increase is evaluated by means of resistance factor. It is the ratio of the alternating current resistance and direct current resistance:

$$k_R = \frac{R_{AC}}{R_{DC}}. \tag{43}$$

The direct consequence of the resistance increase is loss increase; therefore, the resistance factor can be expressed also by the ratio of the losses at alternating current and direct current:

$$k_R = \frac{\Delta P_{AC}}{\Delta P_{DC}}. \tag{44}$$

Loss increase because of skin effect is the reason why it is necessary to deal with this phenomenon in the period of the machine design and parameter determination. Resistance and losses at alternating current can be calculated on the basis of Eq. (43) or (44), if the resistance factor is calculated by means of the equations given below.

In electrical machines the skin effect occurs mainly in the area of the slot but also in the area of end winding. The calculation in these two areas must be made separately, because the magnetic properties of the material in the slot and its vicinity and the end winding are totally different.

Analytical calculation of resistance factor which includes skin effect influence is given in many books dealing with this topic; therefore, here only expressions needed for resistance factor calculation are shown. An important role in the theory of skin effect is the so-called depth of penetration, meaning the depth to which electromagnetic wave penetrates into a material at a given frequency and material conductivity. The depth of penetration depends on the frequency of alternating current, specific electric conductivity of the conductor material σ_c, and vacuum magnetic permeability, because the conductor in the slot is a nonmagnetic material. For example, for cooper at 50 Hz, the depth of penetration yields approximately 1 cm. In **Figure 6a**, it is seen that b_c is the conductor width in the slot and b is the total width of the slot. Then the depth of penetration is:

$$a = \sqrt{\frac{2}{\omega\mu_0\sigma_c}\frac{b}{b_c}}. \tag{45}$$

The conductor height h_c is obviously expressed in ratio to the depth of penetration. Then the conductor height is called the reduced conductor height ξ. It is a dimensionless number:

$$\xi = \frac{h_c}{a} = h_c\sqrt{\frac{\omega\mu_0\sigma_c}{2}\frac{b_c}{b}}. \tag{46}$$

Note that the product of specific electric conductivity and the ratio of conductor width to the slot width express the reduced electric conductivity of the slot area $\sigma_c \frac{b_c}{b}$, because not the whole slot width is filled with the conductor.

If in the slot there are rectangular conductors placed z_a adjacent and z_t on top of each other, the reduced conductor height is calculated according to:

$$\xi = \frac{h_{c0}}{a} = h_{c0}\sqrt{\frac{\omega\mu_0\sigma_c}{2}\frac{z_a b_{c0}}{b}}. \tag{47}$$

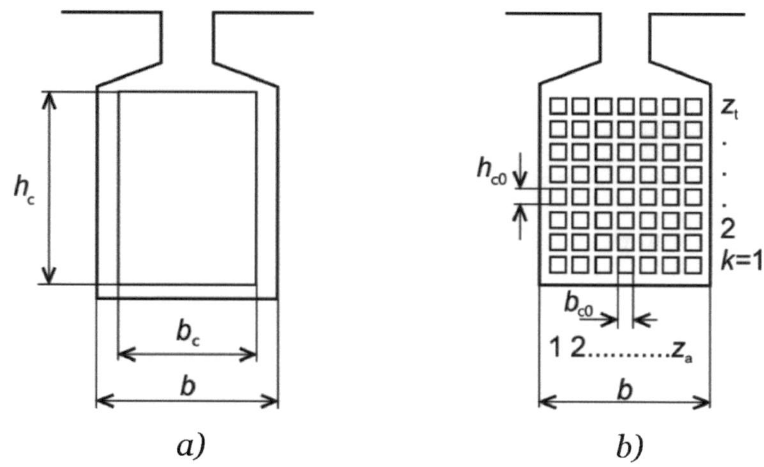

Figure 6.
Determination of reduced conductor height (a) if in the slot with width b *there is only one conductor with the width* b_c *and the height* h_c *(b) if in the slot there are several conductors,* z_a *adjacent conductors and* z_t, *conductors on top of each other. The width of one conductor is* b_{co}, *height* h_{co}.

The resistance factor of the kth layer is:

$$k_{\mathrm{Rk}} = \varphi(\xi) + k(k-1)\psi(\xi), \tag{48}$$

where the functions $\varphi(\xi)$ and $\psi(\xi)$ are derived based on the loss investigation in the conductor placed in the slot of electrical machines and are given as follows [1]:

$$\varphi(\xi) = \xi \frac{sinh\, 2\xi + sin\, 2\xi}{cosh\, 2\xi - cos\, 2\xi}, \tag{49}$$

$$\psi(\xi) = 2\xi \frac{sinh\, \xi - sin\, \xi}{cosh\, \xi + cos\, \xi}. \tag{50}$$

Equation (48) shows that the resistance factor is smallest on the bottom layer and largest on the top layer. This means that in the case of series-connected conductors, the bottommost conductors contribute less to the resistive losses than the topmost conductors. Therefore, it is necessary to calculate the average resistance factor over the slot:

$$k_{\mathrm{Rd}} = \varphi(\xi) + \frac{z_{\mathrm{t}}^2 - 1}{3}\psi(\xi). \tag{51}$$

where z_{t} is number of the conductors on top of each other.
If ξ is in the interval $0 \le \xi \le 1$, Eq. (51) can be simplified:

$$k_{\mathrm{Rd}} = 1 + \frac{z_{\mathrm{t}}^2 - 0.2}{9}\xi^4. \tag{52}$$

Equations above are valid for rectangular conductors. The eddy current losses (skin effect losses) of round wires are 0.59 times the losses of rectangular wire. If in the slot there are round conductors, resistance factor and also the eddy current losses are only 59% of that appeared in the rectangular conductors. Therefore, for the round conductors, Eq. (52) will have a form:

$$k_{\mathrm{Rd}} = 1 + 0.59\frac{z_t^2 - 0.2}{9}\xi^4. \tag{53}$$

An effort of the designer is to reduce the resistance factor what would result in the reduction of the losses. Obviously, it is recommended to divide the height of the conductor: it means to make more layers z_t. As shown in Eq. (52), the resistance factor is proportional to the square of the number of conductors on top of each other z_t, which would look like the resistance factor would increase. But the reduced conductor height is smaller with the smaller conductor height, and according to Eqs. (52) and (53), the reduced conductor height ξ is with exponent 4. Therefore, the resistance factor finally will be lower.

If the conductors are divided into parallel subconductors, which are connected together only at the beginning and at the end of winding, they must be also transposed to achieve effect of the reduction of the resistance factor and of the eddy current (skin effect) losses. Transposition must be made in such a way that all conductors are linked with the equal leakage magnetic field. It means that the changing of the conductor positions must ensure that all conductors engage all possible positions in the slot regarding the leakage magnetic flux. Without transposition of the subconductors, a divided conductor is fairly useless. Further details are given in [1–3].

1.3.2 Influence of skin effect on the winding inductance

If the height of the conductor is large, or if the winding is created only by one bar, as it is in the case of the squirrel cage of asynchronous machines, in the conductors with alternating current, skin effect appears. The skin effect is active also at the medium frequencies and has considerable influence on the resistance (see Section 3.1) and on the winding inductance too. That part of winding, which is positioned on the bottom of the slot, is surrounded by the higher magnetic flux than that on the top of the slot. Therefore, the winding inductance on the top of the slot is lower than that on the bottom of the slot, and therefore the time-varying current is distributed in such a way that the current density on the top of the slot is higher than that on the bottom of the slot. The result is that skin effect increases conductor resistance and reduces slot leakage inductance. To express the decrease of the inductance, the so-called skin effect factor k_L is introduced. This term must be supplemented to the equation for the magnetic permeance. Therefore, Eq. (20), which is valid for the slot on **Figure 2a**, must be corrected to the form:

$$\lambda_{4,L} = k_L \frac{h_4}{3b_4}. \tag{54}$$

To calculate the skin effect factor, we need to define the reduced conductor height:

$$\xi = h_4 \sqrt{\omega \mu_0 \sigma \frac{b_c}{2b_4}}, \tag{55}$$

where b_c is the conductor width in the slot, σ is the specific material conductivity of the conductor, and ω is the angular frequency of the investigated current. For example, in the rotor of induction machine, there is the angular frequency given by the slip and synchronous angular frequency. Then the skin effect factor is a function of this reduced conductor height and the number of conductor layers on top of each other z_t:

$$k_L = \frac{1}{z_t^2}\phi'(\xi) + \frac{z_t^2 - 1}{z_t^2}\psi'(\xi), \tag{56}$$

where

$$\phi'(\xi) = \frac{3}{2\xi}\left(\frac{\sinh 2\xi - \sin 2\xi}{\cosh 2\xi - \cos 2\xi}\right), \tag{57}$$

$$\psi'(\xi) = \frac{1}{\xi}\left(\frac{\sinh \xi + \sin \xi}{\cosh \xi + \cos \xi}\right). \tag{58}$$

In the cage armature, $z_t = 1$; therefore, the skin effect factor is:

$$k_L = \phi'(\xi). \tag{59}$$

In the cage armature, it is usually $h_4 > 2$ cm, and for cooper bars, it is according to Eq. (55) $\xi > 2$. Then $\sinh 2\xi \gg \sin 2\xi$, and $\cosh 2\xi \gg \cos 2\xi$, whereby $\sinh 2\xi \approx \cosh 2\xi$; consequently the k_L is reduced to simple expression:

$$k_L \approx \frac{3}{2\xi}. \tag{60}$$

Example 5. Calculate the slot leakage inductance of aluminum squirrel cage bar $z_Q = 1$ at cold start and 50 Hz supply. The slot shape is according to **Figure 2a**, $b_1 = 0.003$ m, $h_1 = 0.002$ m, $b_4 = 0.008$ m, $h_4 = 0.02$ m, $l' = 0.25$ m, and a slot at height h_4 is fully filled with aluminum bar. The conductivity of aluminum at 20°C is 37 MS/m.

Solution: The permeance factor of that part of slot, which is filled by a conductor without skin effect, is:

$$\lambda_4 = \frac{h_4}{3b_4} = \frac{0.02}{3 \cdot 0.008} = 0.833.$$

The permeance factor of the slot opening is:

$$\lambda_1 = \frac{h_1}{b_1} = \frac{0.002}{0.003} = 0.667.$$

The reduced height ξ of the conductor, which is a dimensionless number, is:

$$\xi = h_4\sqrt{\omega\mu_0\sigma\frac{b_c}{2b_4}} = 0.02\sqrt{2\pi \cdot 50 \cdot 4\pi \cdot 10^{-7} \cdot 37 \cdot 10^6 \frac{0.008}{2 \cdot 0.008}} = 1.71.$$

Then the inductance skin effect factor is:

$$k_L = \frac{1}{z_t^2}\phi'(\xi) + \frac{z_t^2 - 1}{z_t^2}\psi'(\xi) = \phi'(\xi) + \frac{1-1}{1}\psi'(\xi) = \phi'(\xi) = \frac{3}{2\xi}\left(\frac{\sinh 2\xi - \sin 2\xi}{\cosh 2\xi - \cos 2\xi}\right),$$

$$k_L = \frac{3}{2 \cdot 1.71}\left(\frac{\sinh 3.42 - \sin 3.42}{\cosh 3.42 - \cos 3.42}\right) = 0.838,$$

and permeance factor of the slot under the skin effect is:

$$\lambda_d = \lambda_1 + k_L\lambda_4 = 0.667 + 0.838 \cdot 0.833 = 1.37.$$

The leakage inductance of a squirrel cage aluminum bar if skin effect is considered is:

$$L_{d,bar} = \mu_0 l' z_Q^2 \lambda_d = 4\pi \cdot 10^{-7} \cdot 0.25 \cdot 1^2 \cdot 1.37 = 0.43 \cdot 10^{-6}\ \mathrm{H}.$$

Chapter 2

Principles of General Theory of Electrical Machines

2.1 Introduction to the general theory of electrical machines

The theory of individual types of electrical machines from the historical point of view was developed independently. Therefore, also terminology and signing of symbols and subscripts were determined independently. This theory was devoted to the investigation of steady-state conditions and quantities needed for design calculation of electrical machines.

However, the basic principles of electrical machines are based on common physical laws and principles, and therefore a general theory of electrical machines was searched. The first person, who dealt with this topic was Gabriel Kron, who asked the question: "Does a universal arrangement of electrical machine exist from which all known types of electrical machines could be derived by means of simple principles?" An answer to this question resulted in the fact that around the year 1935, G. Kron mathematically formulated general theory and defined universal electrical machine, which at various connections represented most of the known machines.

Kron's theory employed tensor analysis and theory of multidimensional non-Euclidean spaces and therefore was misunderstood and non-employed by majority of the technical engineers. After the year 1950, the first works appeared, in which Kron's theory was simplified and therefore better understood. But only after the personal computers (PC) were spread in a great measure and suitable software facilities were available, the general theory of electrical machines became an excellent working means for investigation of electrical machine properties. Nowadays it represents inevitable equipment of technically educated experts in electrical engineering.

A fundamental feature of the general theory of electrical machines is based on the fact that it generalizes principles and basic equations of all electrical machines on the common base, and in such a way it simplifies their explanation and study. Its big advantage is that it formulates equations of electrical machines in such a form that they are valid in transients as well as in steady-state conditions. In this theory the electrical machines are presented as a system of the stationary and moving mutual magnetically linked electrical circuits, which are defined by the basic parameters: self-winding and mutual winding inductances, winding resistances, and moment of inertia, see [7–9].

The general theory of electrical machines is general in such a sense that it is common for a majority of electrical machines and explains their basic properties and characteristics on the basis of common principles. Further it is applicable for various running conditions: steady state, transients, unsymmetrical, and if they are fed by frequency converter at a non-sinusoidal voltage waveform. On the other side, the electrical machines are idealized by simplifying assumptions.

These simplifying assumptions enable to simplify equations, mainly their solutions. Here are some of them:

a. *The saturation of the magnetic circuit is neglected*. Then the relationship between currents and magnetic fluxes are linear. This assumption is needed to be able to use the principle of the magnetic fluxes' superposition. On the other side, this assumption can have a considerable influence on the correctness of the results. In some cases, this assumption impedes the investigation of some problems, e.g., excitation of the shunt dynamo or running of asynchronous generator in island operation. Then the magnetizing characteristic must be taken into account.

b. *The influence of the temperature on the resistances is neglected.* This assumption can be accepted only in the first approach. If accurate results are needed, which are compared with measurements, it is inevitable to take into account a dependence of the resistances on the temperature.

c. *The influence of the frequency on resistances and inductances is neglected.* In fact, it means that the influence of skin effect and eddy currents is neglected. Again, it is valid that this fact is not involved in the equations of the general theory but at precise calculation this phenomenon is necessary to take into account. It is important mainly in the case of non-sinusoidal feeding from the frequency converters, when higher harmonic components with considerable magnitude appeared.

d. *It is supposed that windings are uniformly distributed around the machine periphery (except concentrating coils of field winding).* In fact, the windings of the real machines are distributed and embedded in many slots, whereby slotting is neglected. In this way the real winding is replaced by current layer on the borderline between the air gap and this part of the machine where the winding is located, and calculation of magnetic fields, inductances, etc. is simplified.

   It should be noted that slotting is not ignored totally. In cooperation with the finite element method (FEM), it is possible to receive waveform of air gap magnetic flux density, where the influence of slotting is clearly seen. It is a distorted waveform for which harmonic analysis must be made and to determine components of the harmonic content. For each harmonic component, the induced voltage can be calculated, and the total induced voltage is given by the sum of all components. In this case the slotting influence is included in the value of the induced voltage.

e. *Winding for alternating currents is distributed sinusoidally.* This assumption means that the real distributed winding with a constant number of the conductors in the slot, with finite number of slots around the machine periphery, is replaced by the winding with conductor (turns) density, varying around the periphery according to the sinusoidal function. This assumption can be used only for winding with many slots and can't be used for concentrating coils of the field windings or for machines with permanent magnets. By this assumption sinusoidal space distribution of the current linkage around the periphery is received, with neglecting of the space harmonic components. In other words, non-sinusoidal waveform of the air gap magnetic flux density induces in such winding only the fundamental voltage component; it means the factor due to winding distribution for all harmonic components is zero.

Next, an arrangement of the universal machine, on the basis of which the general theory was derived, will be given.

2.2 Design arrangement and basic equations of the universal machine in the general theory

A two-pole commutator machine is taken. The theory spread to the multipole arrangement will be carried out if mechanical angles are converted to the electrical angles and mechanical angular speed to the electrical angular speed:

$$\omega = p\Omega, \tag{61}$$

$$\vartheta_{\text{el}} = p\vartheta_{\text{mech}}. \tag{62}$$

The typical phenomenon of the universal machine is that its windings are located in two perpendicular axes to each other: The direct axis is marked "d" and quadrature axis marked q (see **Figure 7**).

Stator has salient poles with one or more windings on the main poles in the d-axis and q-axis. In **Figure 7**, windings f and D are in the d-axis and windings g and Q in the q-axis. These windings can represent field winding (external, shunt, series, according to the connection to the armature), damping, commutating, compensating, and so on (see chapter about the DC machines) or, as we will see later, three-phase winding transformed into the two-axis system d, q.

The rotor's winding with commutator expressed oneself as the winding of the axis that goes through the brushes. If the rotor rotates, conductors of the coils change their position with regard to the stator and brushes, but the currents in the conductors which are located in one pole pitch have always the same direction. In other words, there exists always such conductor which is in a specific position, and the current flows in the given direction.

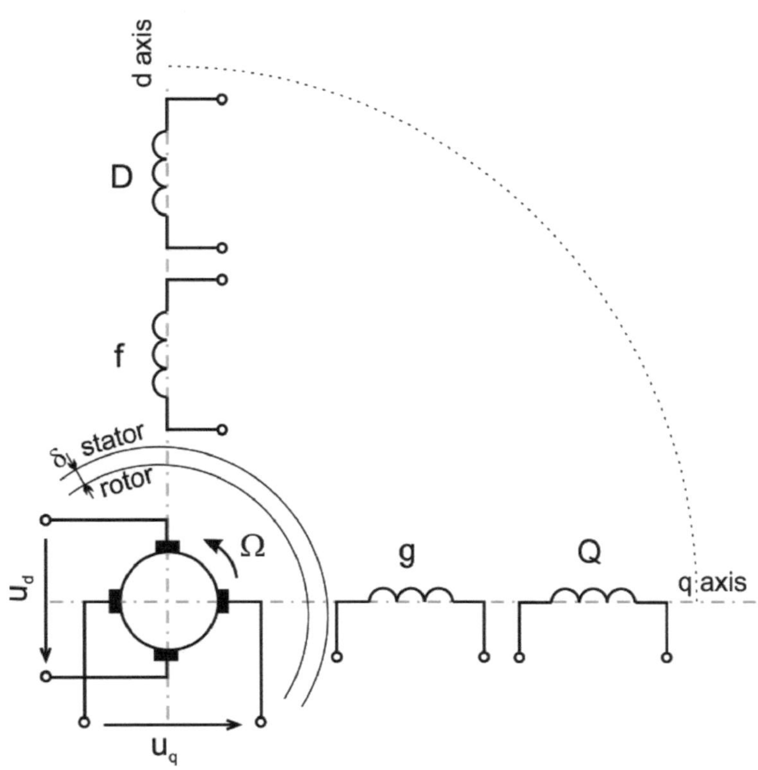

Figure 7.
Design arrangement of the universal machine with marked windings.

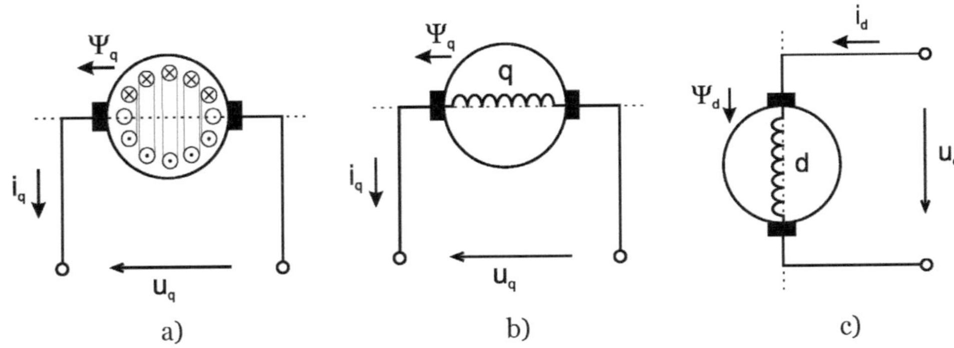

Figure 8.
(a) Replacement of the rotor's winding by solenoid, (b) creation of the quasi-stationary q-winding, and (c) creation of the quasi-stationary d-winding.

Therefore, the rotor's winding with commutator seems to be, from the point of view of magnetic effect, stationary; it means it is quasi-stationary. The magnetic flux created by this winding has always the same direction, given by the link of the given brushes. It is valid for the winding in d-axis and q-axis.

On the commutator, there are two sets of the brushes: one, on terminals of which is voltage u_q, which is located on the q-axis, and the other, on terminals of which is voltage u_d. This voltage is shifted compared with u_q on 90° in the direction of the rotor rotating and is located in the direct d-axis.

2.2.1 Voltage equations in the system dq0

2.2.1.1 Voltage equations of the stator windings

The number of the equations is given by the number of windings. All windings are taken as consumer of the energy. Then the terminal voltage equals the sum of the voltage drops in the windings. The power in the windings is positive; therefore, the voltage and current have the same directions and are also positive. The voltage equations are written according to the 2nd Kirchhoff's law and Faraday's law for each winding. By these equations three basic variables of the given winding, voltage, current, and linkage magnetic flux, are linked. The voltage equation in general for j-th winding, where j = f, D, g, Q, is in the form:

$$u_j = R_j i_j + \frac{d\psi_j}{dt}, \tag{63}$$

where u_j is the terminal voltage, R_j is the resistance, i_j is the current, and ψ_j is the linkage magnetic flux of the j-th winding. For example, for the f-th winding, the equation should be in the form:

$$u_f = R_f i_f + \frac{d\psi_f}{dt}. \tag{64}$$

The linkage magnetic fluxes of the windings are given by the magnetic fluxes created by the currents of the respected winding and those windings which are magnetically linked with them. In general, any winding, including rotors, can be written as:

$$\psi_j = \sum_k \psi_{jk} = \sum_k L_{jk} i_k \text{ where} j,k = f,d,D,q,g,Q. \tag{65}$$

For example, for f-winding, the following is valid:

$$\psi_f = \sum_k \psi_{fk} = \sum_k L_{fk} i_k = L_{ff} i_f + L_{fd} i_d + L_{fD} i_D + L_{fq} i_q + L_{fg} i_g + L_{fQ} i_Q. \quad (66)$$

Equations written in detail for all windings are as follows:

$$\begin{aligned}
\psi_f &= L_{ff} i_f + L_{fd} i_d + L_{fD} i_D + L_{fq} i_q + L_{fg} i_g + L_{fQ} i_Q, \\
\psi_d &= L_{df} i_f + L_{dd} i_d + L_{dD} i_D + L_{dq} i_q + L_{dg} i_g + L_{dQ} i_Q, \\
\psi_D &= L_{Df} i_f + L_{Dd} i_d + L_{DD} i_D + L_{Dq} i_q + L_{Dg} i_g + L_{DQ} i_Q, \\
\psi_q &= L_{qf} i_f + L_{qd} i_d + L_{qD} i_D + L_{qq} i_q + L_{qg} i_g + L_{qQ} i_Q, \\
\psi_g &= L_{gf} i_f + L_{gd} i_d + L_{gD} i_D + L_{gq} i_q + L_{gg} i_g + L_{gQ} i_Q, \\
\psi_Q &= L_{Qf} i_f + L_{Qd} i_d + L_{QD} i_D + L_{Qq} i_q + L_{Qg} i_g + L_{QQ} i_Q.
\end{aligned} \quad (67)$$

In these equations formally written in the order of the windings and their currents, it is shown also, which we already know, that their mutual inductances are zero, because their windings are perpendicular to each other, which results in zero mutual inductance.

2.2.1.2 Voltage equations of the rotor windings

Rotor winding is moving with an angular speed Ω; therefore, not only transformation voltage, which is created by the time varying of the magnetic flux, but also moving (rotating) voltage is induced in it. If the rotating-induced voltage is derived, sinusoidal waveform of the air gap magnetic flux density is assumed.

Rotor winding is composed of two parts, one is located in the d-axis, leading up to the terminals in the d-axis, and the second in the q-axis, leading up to the terminals in the q-axis. In **Figures 9** and **10**, it is shown that not both windings have both voltage components from both linkage magnetic fluxes.

Transformation voltage created by the time variation of ψ_q is induced in the winding in the q-axis, which is with it in magnetic linkage. This flux crosses the whole area of the q-winding turns:

$$u_{trq} = \frac{d\psi_q}{dt}. \quad (68)$$

Linkage magnetic flux ψ_d does not cross the area of any turns of the q-winding; therefore in q-winding there is no induced transformation voltage from ψ_d.

A movement of the q-winding in the marked direction (**Figure 9**) does not cause any rotating induced voltage from ψ_q, because the conductors of the q-winding do not cross magnetic force lines ψ_q; they only move over them.

Rotating voltage in q-winding is induced by crossing the magnetic force lines Φ_d:

$$u_{rotq} = C\Phi_d\Omega, \quad (69)$$

whereby there the known expression from the theory of electrical machines was used:

$$u_i = u_{rotq} = C\Phi_d\Omega = \frac{p}{a}\frac{z}{2\pi}\Phi_d\Omega. \quad (70)$$

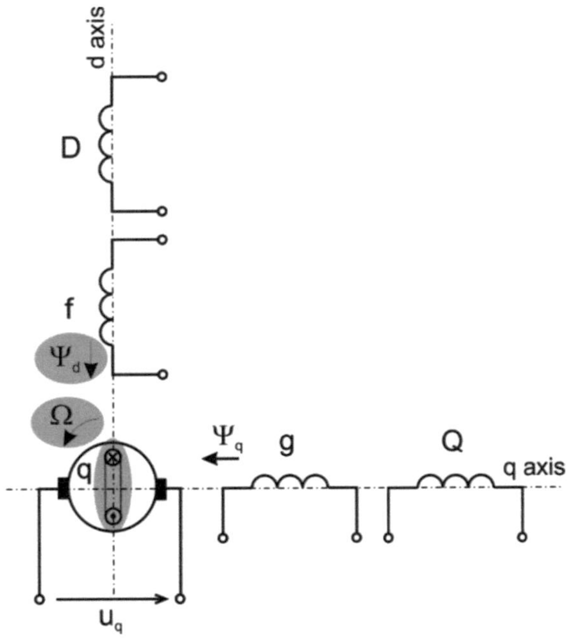

Figure 9.
Illustration of the induced voltage in the q-axis.

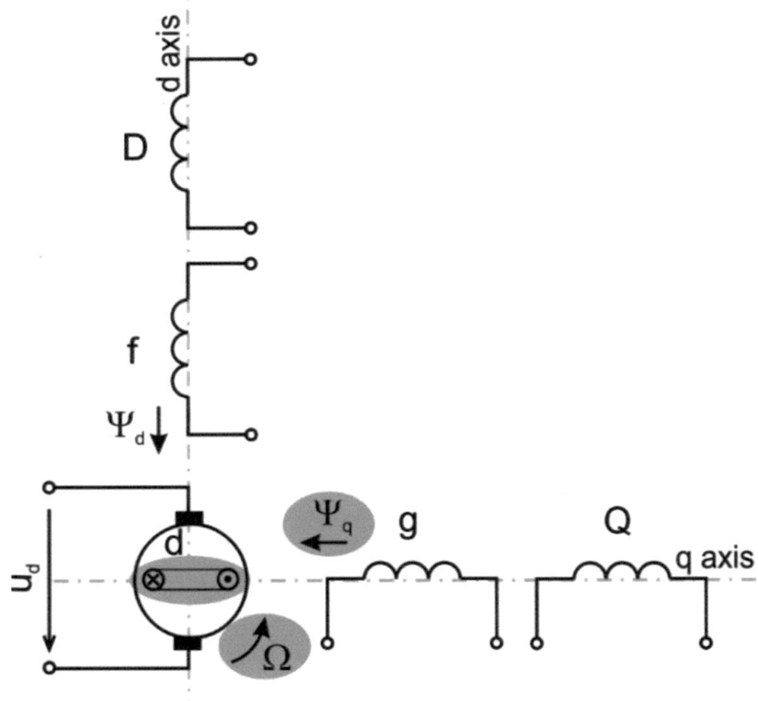

Figure 10.
Illustration of the induced voltage in the d-axis.

This expression can be modified to the generally written mode; it means without regard which axis the winding is, in such a way that it will be written by means of electrical angular speed and linkage magnetic flux.

$$u_{\mathrm{i}} = p\Omega \frac{z}{a2\pi}\Phi = \omega\psi, \tag{71}$$

where half of the conductors z means number of the turns N. Then the linkage magnetic flux includes the next variables:

$$\psi = \frac{z}{a2\pi}\Phi = \frac{N}{a\pi}\Phi = \frac{N}{2a}\frac{2}{\pi}\Phi. \tag{72}$$

In this expression the effective number of turns of one parallel path $N/2a$ is considered, on which the voltage is summed and reduced by the winding factor of the DC machines $2/\pi$. Then the relationship for the induced voltage can be written in general form by means of electrical angular speed and linkage magnetic flux:

$$u_{\mathrm{i}} = p\Omega \frac{z}{a2\pi}\Phi = \omega \frac{N}{2a}\frac{2}{\pi}\Phi. \tag{73}$$

Now the relationship for the terminal voltage in q-axis can be written in the form:

$$u_{\mathrm{q}} = Ri_{\mathrm{q}} + u_{\mathrm{trq}} + u_{\mathrm{rotq}} = Ri_{\mathrm{q}} + \frac{\mathrm{d}\psi_{\mathrm{q}}}{\mathrm{d}t} + \omega\psi_{\mathrm{d}}. \tag{74}$$

As shown in magnetic flux directions, movement and direction of currents in **Figure 9** are in coincidence with the rule of the left hand, meaning for motor operation (consumer). Therefore, all signs in front of the voltages in Eq. (74) are positive.

In d-winding there is induced transformation voltage done by time varying of ψ_{d}, which is in magnetic link with it (this flux crosses the whole area of the d-winding turns). The rotating voltage in d-winding is induced only by crossing magnetic force lines of the ψ_{q} (**Figure 10**). Therefore, the equation for terminal voltage in the d-axis is next:

$$u_{\mathrm{d}} = Ri_{\mathrm{d}} + u_{\mathrm{trd}} - u_{\mathrm{rotd}} = Ri_{\mathrm{d}} + \frac{\mathrm{d}\psi_{\mathrm{d}}}{\mathrm{d}t} - \omega\psi_{\mathrm{q}}. \tag{75}$$

In **Figure 10**, it is seen that current, magnetic flux, and moving directions are in coincidence with the right-hand rule, which is used for generator as a source of electrical energy.

Therefore, the sign in front of the rotating voltage is negative.

2.2.2 Power in the system dq0 and electromagnetic torque of the universal machine

Power and electromagnetic torque of the universal machine will be derived on the basis of energy equilibrium of all windings of the whole machine: We start with the voltage equations of stator and rotor windings, which are multiplied with the appropriate currents and time dt.

For the stator windings, Eq. (63) will be used for the terminal voltage of each winding. This equation will be multiplied by $i_j\mathrm{d}t$ and the result is:

$$u_{\mathrm{j}}i_{\mathrm{j}}\mathrm{d}t = R_{\mathrm{j}}i_{\mathrm{j}}^2\mathrm{d}t + i_{\mathrm{j}}\mathrm{d}\psi_{\mathrm{j}}. \tag{76}$$

For the rotor winding in the d-axis, Eq. (75) multiplied by $i_{\mathrm{d}}\mathrm{d}t$ will be used:

$$u_{\mathrm{d}}i_{\mathrm{d}}\mathrm{d}t = Ri_{\mathrm{d}}^2\mathrm{d}t + i_{\mathrm{d}}\mathrm{d}\psi_{\mathrm{d}} - \omega\psi_{\mathrm{q}}i_{\mathrm{d}}\mathrm{d}t. \tag{77}$$

For the rotor winding in the q-axis, Eq. (74) multiplied by $i_{\mathrm{q}}\mathrm{d}t$ will be used:

$$u_{\mathrm{q}}i_{\mathrm{q}}\mathrm{d}t = Ri_{\mathrm{q}}^2\mathrm{d}t + i_{\mathrm{q}}\mathrm{d}\psi_{\mathrm{q}} + \omega\psi_{\mathrm{d}}i_{\mathrm{q}}\mathrm{d}t. \tag{78}$$

Now the left sides and right sides of these equations are summed, and the result is an equation in which energy components can be identified:

$$\Sigma ui\mathrm{d}t = \Sigma Ri^2\mathrm{d}t + \Sigma i\mathrm{d}\psi + \omega\left(\psi_{\mathrm{d}}i_{\mathrm{q}} - \psi_{\mathrm{q}}i_{\mathrm{d}}\right)\mathrm{d}t. \tag{79}$$

The expression on the left side presents a rise of the delivered energy during the time $\mathrm{d}t$: $\Sigma ui\mathrm{d}t$.

The first expression on the right side presents rise of the energy of the Joule's losses in the windings: $\Sigma Ri^2\mathrm{d}t$.

The second expression on the right side is an increase of the field energy: $\Sigma i\mathrm{d}\psi$.

The last expression means a rise of the energy conversion from electrical to mechanical form in the case of motor or from mechanical to electrical form in the case of generator: $\omega\left(\psi_{\mathrm{d}}i_{\mathrm{q}} - \psi_{\mathrm{q}}i_{\mathrm{d}}\right)\mathrm{d}t$.

The instantaneous value of the electromagnetic power of the converted energy can be gained if the expression for energy conversion will be divided by time $\mathrm{d}t$:

$$p_{\mathrm{e}} = \frac{\omega\left(\psi_{\mathrm{d}}i_{\mathrm{q}} - \psi_{\mathrm{q}}i_{\mathrm{d}}\right)\mathrm{d}t}{\mathrm{d}t} = \omega\left(\psi_{\mathrm{d}}i_{\mathrm{q}} - \psi_{\mathrm{q}}i_{\mathrm{d}}\right) = p\Omega\left(\psi_{\mathrm{d}}i_{\mathrm{q}} - \psi_{\mathrm{q}}i_{\mathrm{d}}\right), \tag{80}$$

where p is the number of pole pairs. The subscript "e" is used to express "electromagnetic power p_{e}," i.e., air gap power, where also the development of instantaneous value of electromagnetic torque t_{e} is investigated:

$$p_{\mathrm{e}} = t_{\mathrm{e}}\Omega. \tag{81}$$

If left and right sides of Eqs. (80) and (81) are put equal, an expression for the instantaneous value of the electromagnetic torque in general theory of electrical machines yields:

$$t_{\mathrm{e}} = p\left(\psi_{\mathrm{d}}i_{\mathrm{q}} - \psi_{\mathrm{q}}i_{\mathrm{d}}\right). \tag{82}$$

If the motoring operation is analyzed, it can be seen that at known values of the terminal voltages (six equations) and known parameters of the windings, there are seven unknown variables, because except six currents in six windings there is also angular rotating speed, which is an unknown variable. Therefore, further equation must be added to the system. It is the equation for mechanical variables:

$$m_{\mathrm{e}} = J\frac{\mathrm{d}\Omega}{\mathrm{d}t} + t_{\mathrm{L}}, \tag{83}$$

in which it is expressed that developed electromagnetic torque given by Eq. (82) covers not only the energy of the rotating masses $J\frac{\mathrm{d}\Omega}{\mathrm{d}t}$ with the moment of inertia J but also the load torque t_{L}.

Therefore, from the last two equations, the time varying of the mechanical angular speed can be calculated:

$$\frac{d\Omega}{dt} = \frac{1}{J}(t_e - t_L) = \frac{1}{J}\left[p\left(\psi_d i_q - \psi_q i_d\right) - t_L\right]. \tag{84}$$

For the time varying of the electrical angular speed, which is directly linked with the voltage equations, we get:

$$\frac{d\omega}{dt} = \frac{p}{J}\left[p\left(\psi_d i_q - \psi_q i_d\right) - t_L\right]. \tag{85}$$

These equations will be simulated if transients of electrical machines are investigated.

2.3 Application of the general theory onto DC machines

If the equivalent circuit of the universal machine and equivalent circuits of the DC machines are compared in great detail, it can be seen that the basic principle of the winding arrangement in two perpendicular axes is very well kept in DC machines. It is possible to find a coincidence between generally defined windings f, D, Q, g, d, and q and concrete windings of DC machines, e.g., in this way:

The winding “f” represents field winding of DC machine.

The winding “D” either can represent series field winding in the case of compound machines, whereby the winding “f” is its shunt field winding, or, if it is short circuited, can represent damping effects during transients of the massive iron material of the machines. However, it is true that to investigate the parameters of such winding is very difficult [1].

The windings “g” and “Q” can represent stator windings, which are connected in series with the armature winding, if they exist in the machine. They can be commutating pole winding and compensating winding.

Windings “d” and “q” are the winding of the armature, but in the case of the classical construction of DC machine, where there is only one pair of terminals, and eventually one pair of the brushes in a two-pole machine, only q-winding and terminals with terminal voltage u_q will be taken into account. The winding in d-axis will be omitted, and by this way also terminals in d-axis, its voltage u_d, and current i_d will be cancelled.

The modified equivalent circuit of the universal electrical machine applied to DC machine is in **Figure 11**.

2.3.1 Separately excited DC machine

The field winding of the separately excited DC machine is fed by external source of DC voltage and is not connected to the armature (see **Figure 12**). Let us shortly explain how the directions of voltages, currents, speed, and torques are drawn: The arrowhead of the induced voltage is moving to harmonize with the direction of the magnetic flux in the field circuit. The direction of this movement means the direction of rotation and of developed electromagnetic (internal) torque. The load torque and the loss torque (the torque covering losses) are in opposite directions. The source of voltage is on the terminals and current flows in the opposite direction. On the armature there are arrowheads of voltage and current in coincidence, because the armature is a consumer.

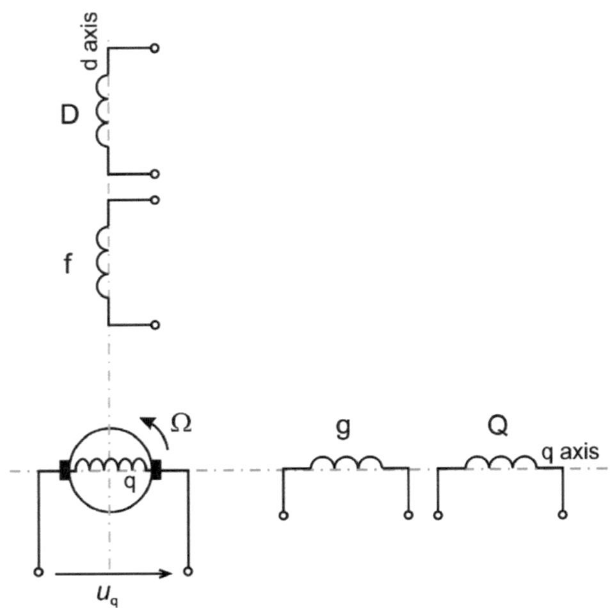

Figure 11.
Modified equivalent circuit of the universal machine applied on DC machine.

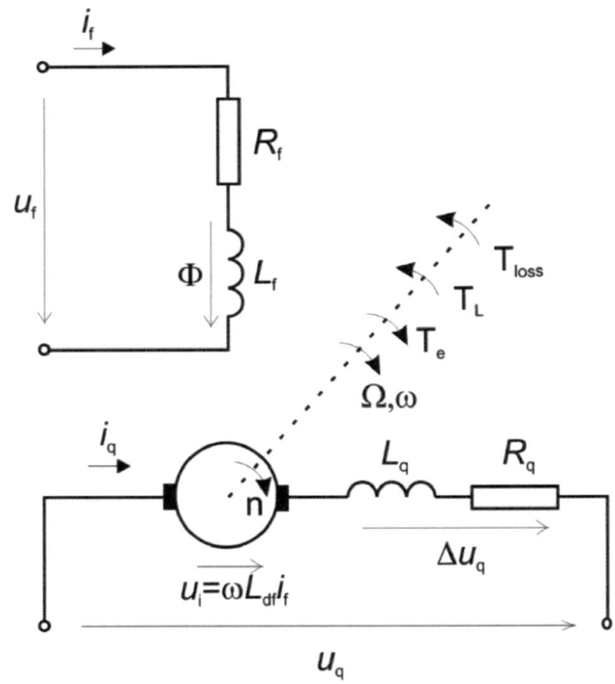

Figure 12.
Equivalent circuit of DC machine with separate excitation in motoring operation.

To solve transients' phenomena, a system of the voltage equations of all windings is needed (for simplification g-winding is omitted):

$$u_f = R_f i_f + \frac{d\psi_f}{dt}, \tag{86}$$

$$u_{\mathrm{D}} = R_{\mathrm{D}} i_{\mathrm{D}} + \frac{\mathrm{d}\psi_{\mathrm{D}}}{\mathrm{d}t}, \tag{87}$$

$$u_{\mathrm{Q}} = R_{\mathrm{Q}} i_{\mathrm{Q}} + \frac{\mathrm{d}\psi_{\mathrm{Q}}}{\mathrm{d}t}, \tag{88}$$

$$u_{\mathrm{q}} = R_{\mathrm{q}} i_{\mathrm{q}} + \frac{\mathrm{d}\psi_{\mathrm{q}}}{\mathrm{d}t} + \omega\psi_{\mathrm{d}}, \tag{89}$$

where:

$$\psi_{\mathrm{f}} = L_{\mathrm{ff}} i_{\mathrm{f}} + L_{\mathrm{fD}} i_{\mathrm{D}}, \tag{90}$$

$$\psi_{\mathrm{D}} = L_{\mathrm{DD}} i_{\mathrm{D}} + L_{\mathrm{Df}} i_{\mathrm{f}}, \tag{91}$$

$$\psi_{\mathrm{Q}} = L_{\mathrm{QQ}} i_{\mathrm{Q}} + L_{\mathrm{Qq}} i_{\mathrm{q}}, \tag{92}$$

$$\psi_{\mathrm{q}} = L_{\mathrm{qq}} i_{\mathrm{q}} + L_{\mathrm{qQ}} i_{\mathrm{Q}}, \tag{93}$$

$$\psi_{\mathrm{d}} = L_{\mathrm{dd}} i_{\mathrm{d}} + L_{\mathrm{df}} i_{\mathrm{f}} + L_{\mathrm{dD}} i_{\mathrm{D}} = L_{\mathrm{df}} i_{\mathrm{f}} + L_{\mathrm{dD}} i_{\mathrm{D}}, \tag{94}$$

because the current in “d”-winding is zero, seeing that d-winding is omitted. In addition, the fact that mutual inductance of two perpendicular windings is zero was considered.

Then equation for electromagnetic torque is needed. This equation shows that electromagnetic torque in DC machine is developed in the form (again the member with the current i_{d} is cancelled):

$$t_{\mathrm{e}} = p\left(\psi_{\mathrm{d}} i_{\mathrm{q}} - \psi_{\mathrm{q}} i_{\mathrm{d}}\right) = p\psi_{\mathrm{d}} i_{\mathrm{q}}, \tag{95}$$

and that it covers the energy of the rotating mass given by moment of inertia, time varying of the mechanical angular speed, and load torque:

$$t_{\mathrm{e}} = J\frac{\mathrm{d}\Omega}{\mathrm{d}t} + t_{\mathrm{L}}. \tag{96}$$

A checking of equation for electromagnetic torque of DC machines for steady-state conditions will be done if for ψ_{d}, Eq. (72) is applied to the d-axis:

$$\psi_{\mathrm{d}} = \frac{N}{2a}\frac{2}{\pi}\Phi_{\mathrm{d}} \tag{97}$$

is introduced to Eq. (95) for the torque, whereby for the current the subscript “a” is employed and used for the armature winding and the number of the conductors z is taken as double number of the turns N:

$$T_{\mathrm{e}} = p\psi_{\mathrm{d}} i_{\mathrm{q}} = p\frac{N}{2a}\frac{2}{\pi}\Phi_{\mathrm{d}} I_{\mathrm{a}} = \frac{p}{a}\frac{z}{2\pi}\Phi_{\mathrm{d}} I_{\mathrm{a}} = C\Phi_{\mathrm{d}} I_{\mathrm{a}}. \tag{98}$$

If the damping winding D, neither the windings in the quadrature axis Q, g, are not taken into account and respecting Eq. (90) for linkage magnetic flux, we get equations as they are presented below. The simplest system of the voltage equations is as follows:

$$u_{\mathrm{f}} = R_{\mathrm{f}} i_{\mathrm{f}} + L_{\mathrm{f}}\frac{\mathrm{d}i_{\mathrm{f}}}{\mathrm{d}t}, \tag{99}$$

$$u_q = R_q i_q + L_q \frac{di_q}{dt} + \omega \psi_d = R_q i_q + L_q \frac{di_q}{dt} + \omega L_{df} i_f. \quad (100)$$

The expression $\psi_d = L_{df} i_f$ shows that the linkage flux ψ_d in Eq. (94) is created by the mutual inductance L_{df} between the field winding and armature winding (by that winding which exists there and is brought to the terminals through the brushes in the q-axis).

We get from Eq. (96) the equation for calculation of the time varying of the mechanical angular speed:

$$\frac{d\Omega}{dt} = \frac{1}{J}(p\psi_d i_q - t_L) = \frac{1}{J}(pL_{df} i_f i_q - t_L), \quad (101)$$

and the electrical angular speed ω, which appears in the voltage equations, is valid:

$$\frac{d\omega}{dt} = \frac{p}{J}(p\psi_d i_q - t_L) = \frac{p}{J}(pL_{df} i_f i_q - t_L). \quad (102)$$

In this way, a system of three equations (Eqs. (99), (100), and (102)), describing the smallest number of windings (three), was created. The solution of these equations brings time waveforms of the unknown variables ($i_q = f(t)$, $i_f = f(t)$, and $\omega = f(t)$).

2.3.1.1 Separately excited DC motor

If a DC machine is in motoring operation, the known variables are terminal voltages, moment of inertia, load torque, and parameters of the motor, i.e., resistances and inductances of the windings.

Unknown variables are currents, electromagnetic torque, and angular speed. Therefore, Eqs. (99) and (100) must be adjusted for the calculation of the currents:

$$\frac{di_f}{dt} = \frac{1}{L_f}(u_f - R_f i_f), \quad (103)$$

$$\frac{di_q}{dt} = \frac{1}{L_q}(u_q - R_q i_q - \omega \psi_d). \quad (104)$$

The third equation is Eq. (102). It is necessary to solve these three equations, Eqs. (102)–(104), to get time waveforms of the unknown field current, armature current, and electrical angular speed, which can be recalculated to the mechanical angular speed or revolutions per minute: $\Omega = \omega/p$ or $n = 60\Omega/2\pi$ min^{-1}, at the known terminal voltages and parameters of the motor.

As it was seen, a very important part of the transients' simulations is determination of the machine parameters, mainly resistances and inductances but also moment of inertia. The parameters can be calculated in the process of the design of electrical machine, as it was shown in Chapter 1. However, the parameters can be also measured if the machine is fabricated. A guide how to do it is given in [8]. The gained parameters are introduced in equations, and by means of simulation programs, the time waveforms are received. After the decay of the transients, the variables are stabilized; it means a steady-state condition occurs. The simulated waveforms during the transients can be verified by an oscilloscope and steady-state conditions also by classical measurements in steady state.

Designers in the process of the machine design can calculate parameters on the basis of geometrical dimensions, details of construction, and material properties. If they use the above derived equations, they can predestine the properties of the designed machine in transients and steady-state conditions. This is a very good method on how to optimize machine construction in a prefabricated period. When the machine is manufactured, it is possible to verify the parameters and properties by measurements and confirm them or to make some corrections.

2.3.1.2 Simulations of the concrete separately excited DC motor

The derived equations were applied to a concrete motor, the data of which are in **Table 1**. The fact that the motor must be fully excited before or simultaneously with applying the voltage to the armature must be taken into account. Demonstration of the simulation outputs is in **Figure 13**. In **Figure 13a–d**, time waveforms of the simulated variables $i_f = f(t)$, $i_q = f(t)$, $n = f(t)$, and $t_e = f(t)$ are shown after the voltage is applied to the terminals of the field winding in the instant of $t = 0.1$ s. After the field current i_f is stabilized, at the instant $t = 0.6$ s, the voltage was applied to the armature terminals. After the starting up, the no-load condition happened, and the rated load was applied at the instant $t = 1$ s.

In **Figure 13e–g**, basic characteristics of $n = f(T_e)$ are shown for the steady-state conditions. They illustrate methods on how the steady-state speed can be controlled: by controlling the armature terminal voltage U_q, by resistance in the armature circuit R_q, as well as by varying the field current i_f.

Figure 13h points to the fact that value of the armature current I_{qk} does not depend on the value of the field current i_f and also shows the typical feature of the motor with the rigid mechanical curve that the feeding armature current I_q is very high if motor is stationary; it means such motor has a high short circuit current. This is the reason why the speed control is suitable to check value of the feeding current, which can be ensured by the current control loop.

2.3.1.3 Separately excited generator

Equations for universal machine are derived in general; therefore, they can be used also for generating operation. If the prime mover is taken as a source of stiff speed, then the time changing of the speed can be neglected, i.e., $d\Omega/dt = 0$, and Ω = const is taken. In addition, the current in the armature will be reversed, because now the induced voltage in the armature is a source for the whole circuit (see **Figure 14**). According to Eq. (96), equilibrium occurs between the driving torque of prime mover T_{hn} and electromagnetic torque T_e, which act against each other, i.e., the prime mover is loaded by the developed electromagnetic torque. If

$U_{qN} = U_{fN} = 84$ V (in motoring)	$R_q = 0.033\ \Omega$
$I_{qN} = 220$ A	$L_q = 0.324$ mH
$I_{fN} = 6.4$ A	$R_f = 13.2\ \Omega$
$n_N = 3200\ \text{min}^{-1}$	$L_f = 1.5246$ H
$T_N = 48$ Nm	$L_{qf} = 0.0353$ H
$P_N = 16$ kW	$J = 0.04$ kg m^2
$p = 1$	$T_{e0} = 0.2$ Nm

Table 1.
Nameplate and parameters of the simulated separately excited DC motor.

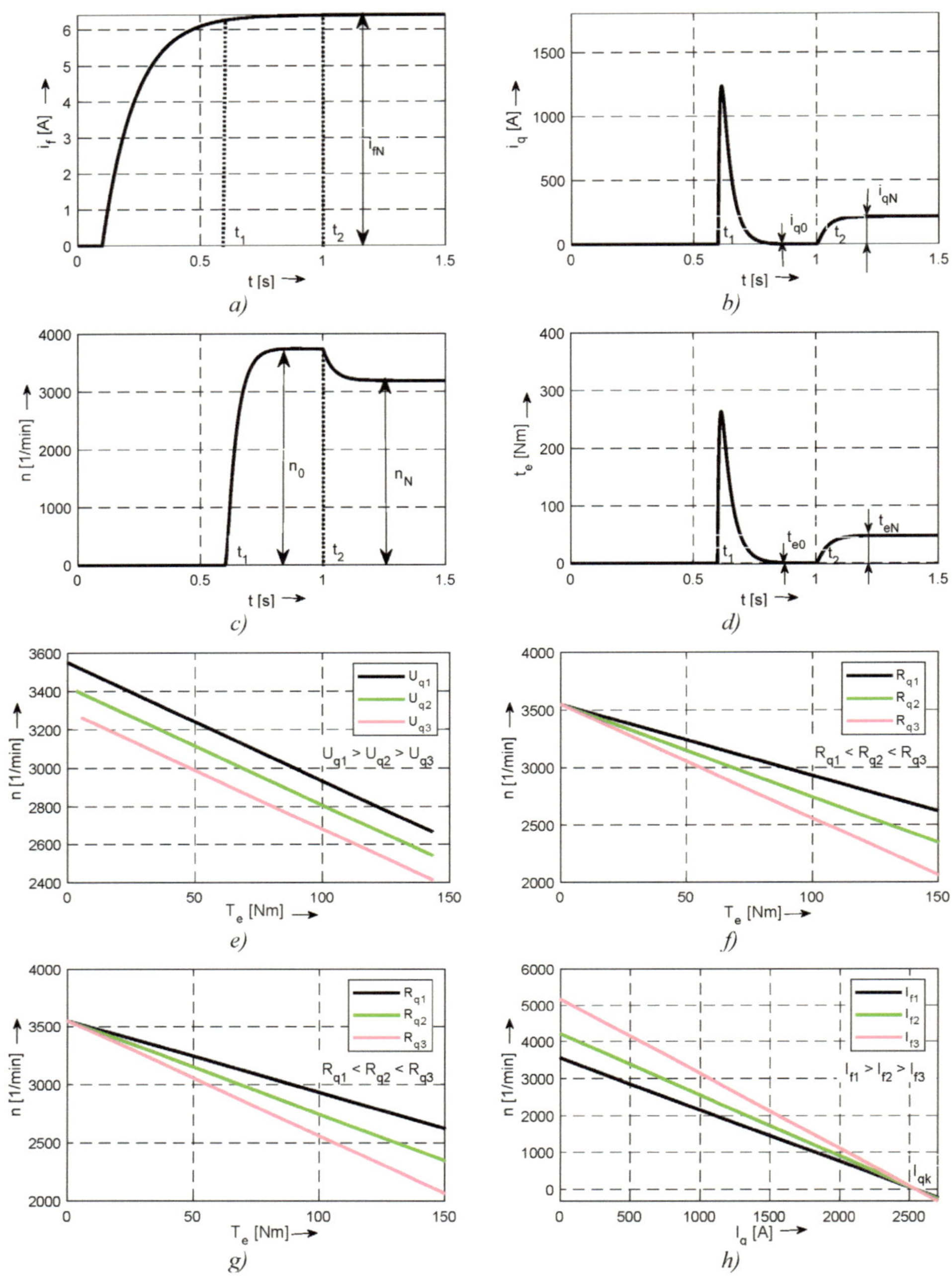

Figure 13.
Simulation waveforms of the separately excited DC motor. Time waveforms of (a) field current, (b) armature current, (c) speed, and (d) developed electromagnetic torque. Dependence of the speed on the torque in steady state, for (e) various terminal voltages, (f) various resistances connected in series with armature, (g) various field currents, and (h) dependence of speed on the armature current for the various field currents.

the analysis is very detailed, it is possible to define dependence of the angular speed of the prime mover on the load by specific function according the mechanical characteristic and to introduce this function into the equation of the torque equilibrium.

The constant speed of the prime mover equations will be changed in comparison with motoring operation. The armature current is in the opposite direction, because now the induced voltage in the armature is a source and current flows from the

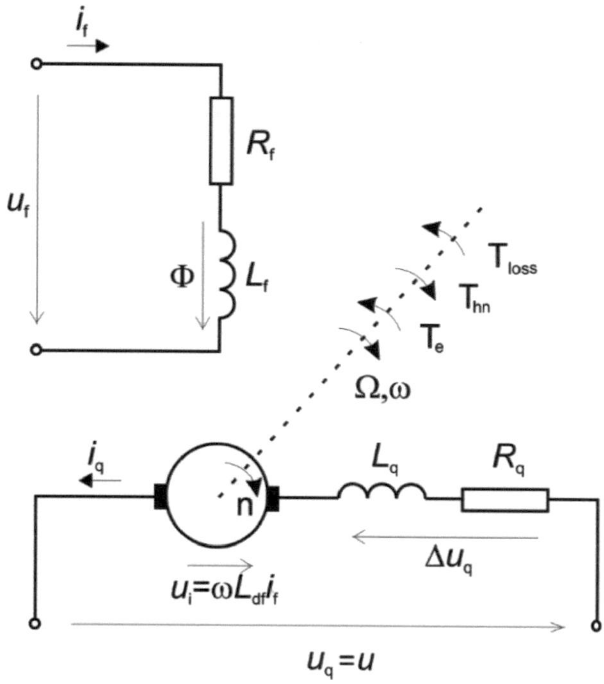

Figure 14.
Equivalent circuit of the separately excited DC generator. The driving torque (T_{hn}) in generating operation is delivered by prime mover.

source. An electrical load is connected to the terminals; therefore the voltage and current on the load are in the same directions. The induced voltage is divided between voltage drops on the resistances and inductances of the winding and on the terminal voltage. Terminal voltage is given by the resistance of the load and its current. Equations are created in this sense:

$$u_i = \omega \psi_d = \omega L_{df} i_f = u + R_q i_q + L_q \frac{di_q}{dt}, \tag{105}$$

and simultaneously the terminal voltage is given by equation of the load:

$$u = R_L i_q. \tag{106}$$

2.3.1.4 Simulations of a concrete separately excited DC generator

An electrical machine, the data of which are given in **Table 1**, is used also for simulation in generating operation. The dynamo is kept at constant speed and is fully excited before any loading occurs.

Simulation waveforms in **Figure 15a–e** show time dependence of the variables: i_f = f(t), u_i = f(t) i_q = f(t), u_q=f(t), and t_e = f(t).

Dynamo is rotating by the rated speed, and at the time t = 0.1 s is excited by the rated field voltage. After the field current is stabilized, at the instant t = 1 s, the dynamo is loaded by the rated current. In **Figure 15f**, there is a waveform of the armature voltage versus armature current u_q = f(i_q). It is the so-called stiff voltage characteristic, i.e., at the big change of the load, the voltage is almost constant. Its moderate fall is caused by voltage drops in the area of the rated load and by armature reaction.

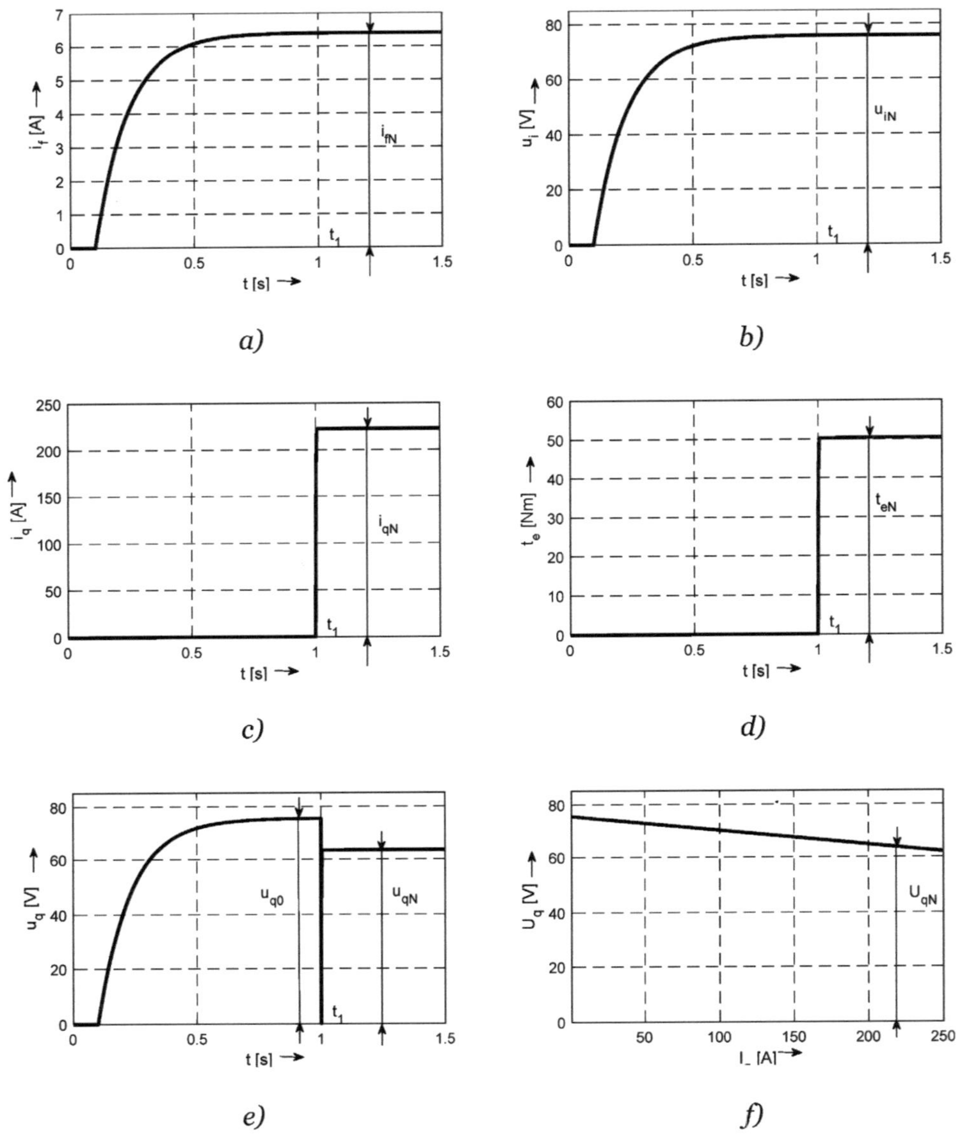

Figure 15.
Simulation of the separately excited dynamo: time waveforms of (a) field current, (b) induced voltage, (c) armature current, (d) developed electromagnetic torque, (e) terminal voltage, and (f) dependence of the terminal voltage on the load current in steady-state conditions (external characteristic).

2.3.2 Shunt wound DC machine

This machine is so called because the field circuit branch is in shunt, i.e., parallel, with that of the armature. **Figure 16** shows equivalent circuits of the shunt machines, in motoring and generating operation. As it is seen, the shunt motor differs from the separately excited motor because the shunt motor has a common source of electrical energy for armature as well as for field winding. Therefore, the field winding is connected parallel to the armature, which results in the changing of equations. In Eqs. (99) and (100), the terminal voltages in both windings are identical:

$$u_f = R_f i_f + L_f \frac{di_f}{dt} = u_q = u, \tag{107}$$

$$u_{\mathrm{q}} = R_{\mathrm{q}} i_{\mathrm{q}} + L_{\mathrm{q}} \frac{\mathrm{d} i_{\mathrm{q}}}{\mathrm{d} t} + \omega \psi_{\mathrm{d}} \tag{108}$$

The power input is given by the product of terminal voltage and the total current i, which is a sum of the currents in both circuits:

$$i = i_{\mathrm{q}} + i_{\mathrm{f}}. \tag{109}$$

The power output is given by the load torque on the shaft and the angular speed. The developed electromagnetic torque is given by equation:

$$t_{\mathrm{e}} = p\left(\psi_{\mathrm{d}} i_{\mathrm{q}} - \psi_{\mathrm{q}} i_{\mathrm{d}}\right) = p \psi_{\mathrm{d}} i_{\mathrm{q}} = p L_{\mathrm{df}} i_{\mathrm{f}} i_{\mathrm{q}}. \tag{110}$$

The time waveform of the electrical angular speed is given by Eq. (102).

2.3.2.1 Simulations of the concrete DC shunt motor

To get simulations of DC shunt motor transients, it is necessary to solve equations from Eq. (107) to Eq. (110) and Eq. (102). Terminal voltage and parameters are known; currents and speed time waveforms are unknown (see **Figure 17**). Here the investigated motor has the same data as they are in **Table 1**.

Because this motor reaches its rated field current at the same field voltage u_{f} as it is the armature voltage u_{q} (at u_{q} = 84 V, the field current waveforms reaches the value of i_{f} = 6.4 A), the simulated time waveforms do not differ from the waveforms of the separately excited DC motor (**Figure 17a–c**). At the other waveforms (**Figures 17d–f**), there are some differences, e.g., variation of the terminal voltage influences not only armature current but also the field current. This fact results in the almost constant speed if terminal voltage is changed. It is proven by the waveforms in **Figure 17d**, which show that in the region till 50 Nm, i.e., rated torque T_{N}, there is no changing of the speed, even in no load condition. Therefore, this kind of speed control is not employed.

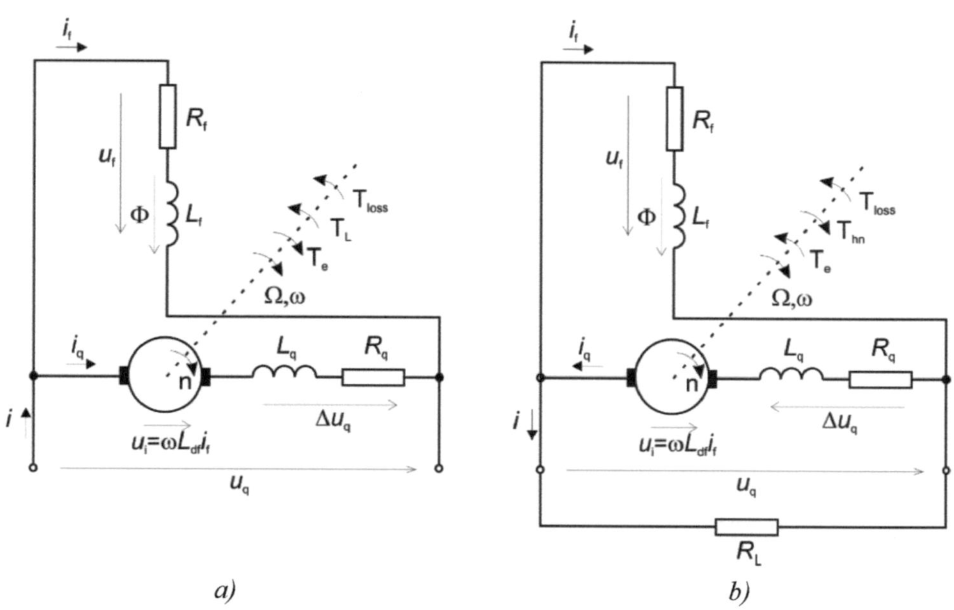

Figure 16.
Equivalent circuits of the shunt DC machine in (a) motoring and (b) generating operation.

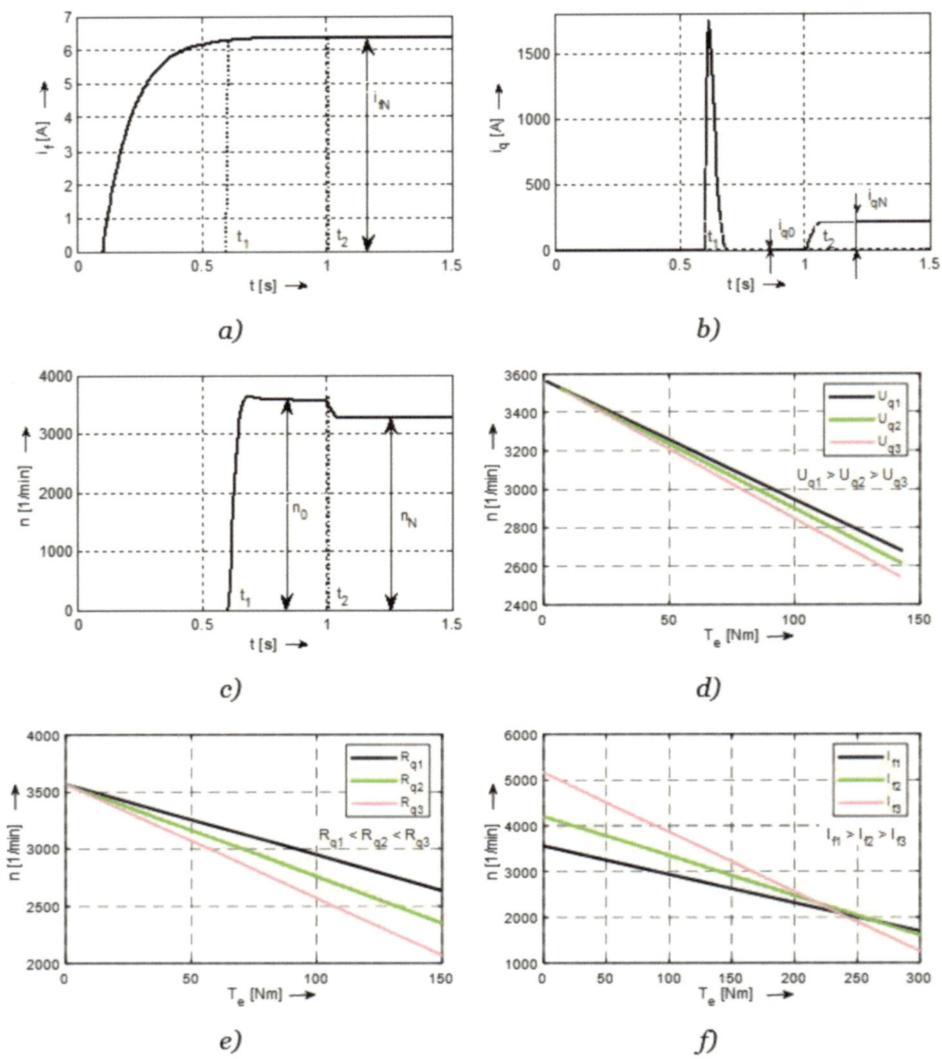

Figure 17.
Simulations of the DC shunt motor. Time waveforms of the (a) field current, (b) armature current, (c) speed, and (d) speed vs. torque for various terminal voltages and further in steady-state conditions waveforms of the (e) speed vs. torque for various rheostats connected in series with armature and (f) speed vs. torque for various field currents at constant terminal voltage.

The control of the field current is carried out by variation of the field rheostat, which ensures decreasing of the field current i_f at the constant field and terminal voltage u_q.

2.3.2.2 Shunt generator

A shut generator (dynamo) differs from the separately excited dynamo by an essential way because a source for the field current is its own armature, where a voltage must be at first induced. To ensure this, some conditions must be filled. They are as follows: (1) some residual magnetism must exist in the magnetic system of the stator, which enables building up of the remanent voltage, if dynamo rotates, (2) resistance in the field circuit must be smaller than a critical resistance, (3) speed must be higher than a critical speed, and (4) there must be correct direction of rotation and connection between polarity of the excitation and polarity of induced voltage in the armature.

Because the field current depends on the terminal voltage, and this terminal voltage on the induced voltage, which again depends on the field current, this mutual dependence must be taken into account in simulations by magnetizing curve of the investigated machine, i.e., induced voltage vs. field current $U_0 = U_i = f(I_f)$, which can be measured. The measurement of this curve can be made only with separate excitation. The speed of the prime mover is taken constant.

Equation (107) is valid, but Eq. (108) is changed, because the terminal voltage is smaller than induced voltage because of the voltage drops, or opposite, induced voltage covers terminal voltage as well as voltage drops:

$$u_i = \omega\psi_d = \omega L_{df} i_f = u + R_q i_q + L_q \frac{di_q}{dt} \tag{111}$$

and armature current supplies field circuit as well as load circuit. Then the load current is:

$$i = i_q - i_f, \tag{112}$$

whereby the terminal voltage is given by the load current and load resistance:

$$u = u_q = R_L i. \tag{113}$$

2.3.2.3 Simulation of a shunt DC dynamo

A machine, in which its data are in **Table 1**, was used for simulations of transients and steady-state conditions. In addition it is necessary to measure magnetizing curve $U_i = f(I_f)$, which is shown in **Figure 18** for the investigated machine.

During the simulation the machine is kept on the constant speed, and simulation starts with the connection of the armature to the field circuit. Because of remanent magnetic flux, in the armature there is induced small remanent voltage U_{irem}, which pushes through the armature circuit and field circuit small field current, by which the magnetic flux and induced voltage will be increased. This results in higher field current and higher induced voltage, which is gradually increased until it reaches the value of the induced voltage in no load condition u_{io}. Simulation waveforms are in

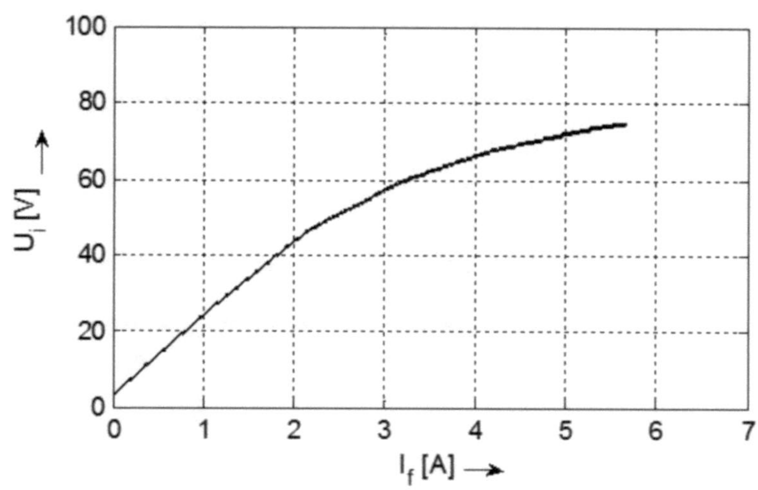

Figure 18.
Magnetizing curve $U_i = f(I_f)$ *for investigated machine.*

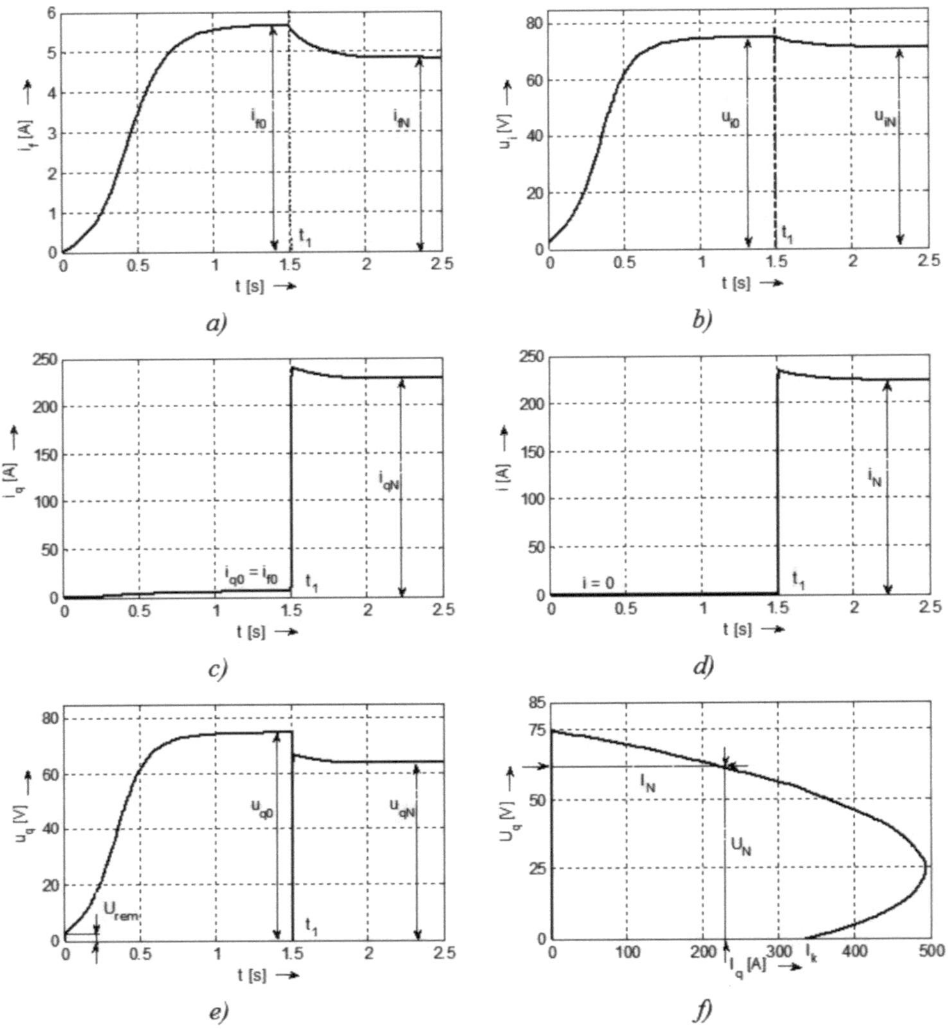

Figure 19.
Simulations of the shunt dynamo: time waveforms of (a) field current, (b) induced voltage, (c) armature current, (d) load current, (e) terminal voltage, and (f) terminal voltage vs. load current in steady-state conditions at the constant resistance in the field circuit when the load resistance is changed.

Figure 19a–e. They show the time waveforms of variables $i_f = f(t)$, $u_i = f(t)$, $i_q = f(t)$, $i = f(t)$, and $u_q = f(t)$ from the instant of connecting until the transients are in a steady-state condition in the time of $t = 1.5$ s.

A waveform of $U_q = f(I_q)$ is shown in **Figure 19f**. It is terminal voltage U_q vs. load current I_q. As it was mentioned, it is a basic characteristic for all sources of electrical energy, and in the case of shunt dynamo, it is seen that there is also a stiff characteristic, similar to the case of the separately excited dynamo but only till the rated load. In addition, it is immune to the short circuit condition, because short circuit current can be smaller than its rated current I_N. This performance is welcomed in the applications where this feature was required, e.g., in cars, welding set, etc.

2.3.3 DC series machine

A DC series machine has its field winding connected in series with its armature circuit, as it is seen in **Figure 20** for motoring and generating operation.

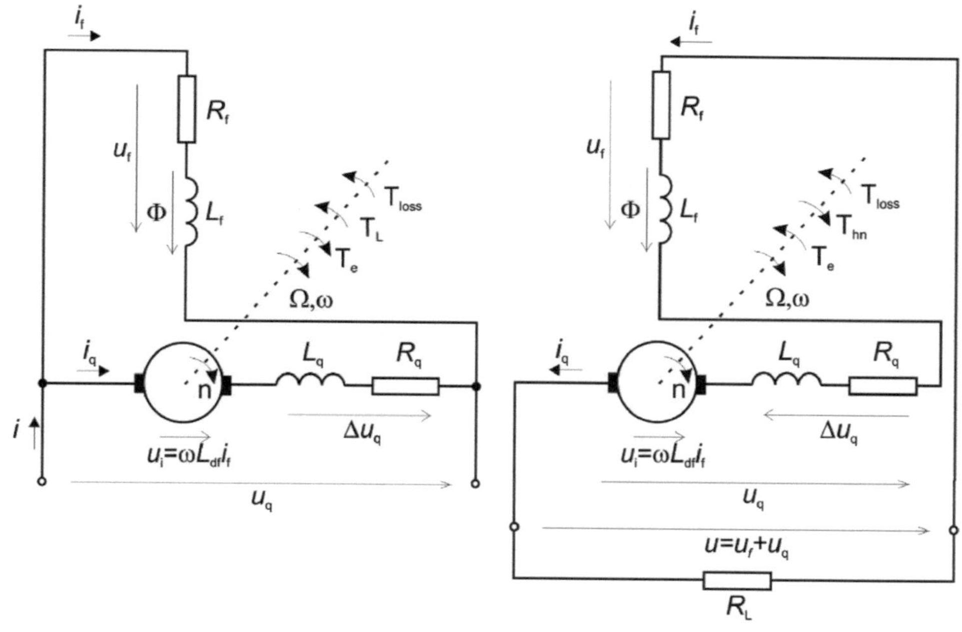

Figure 20.
Equivalent circuits of the series machine (a) in motoring and (b) in generating operation.

This connection essentially influences properties and shapes of characteristics of the series machine and also equations needed for investigations of its properties.

2.3.3.1 Series DC motor

For a series DC motor, it is typical that the terminal voltage is a sum of the voltages in the field circuit and in the armature circuit:

$$u = u_q + u_f = R_q i_q + L_q \frac{di_q}{dt} + \omega L_{df} i_f + R_f i_f + L_f \frac{d\,i_f}{dt}, \tag{114}$$

but because of only one current flowing in the whole series circuit, the next is valid:

$$i = i_q = i_f \tag{115}$$

and Eq. (114) is simplified:

$$u = u_q + u_f = \left(R_q + R_f\right)i + \left(L_q + L_f\right)\frac{di}{dt} + \omega L_{df} i. \tag{116}$$

Equation (110) for electromagnetic torque is also changed because of only one current:

$$t_e = p\left(\psi_d i_q - \psi_q i_d\right) = p\psi_d i = pL_{df} i^2 \tag{117}$$

and angular speed is gained on the basis of the equation:

$$t_e = pL_{df} i^2 = J\frac{d\Omega}{dt} + t_L = \frac{J}{p}\frac{d\omega}{dt} + t_L. \tag{118}$$

2.3.3.2 Simulations of a DC series motor

The time waveforms of the current, developed electromagnetic torque and angular speed, which can be recalculated to the revolutions per minute, are based on Eqs. (116)–(118). In **Figure 20**, there are simulated waveforms of the motor; the data of which are shown in **Table 2**.

Simulated waveforms in **Figure 21a–f** show time waveforms of the variables $i_f = i_q = f(t)$, $t_e = f(t)$, and $n = f(t)$ after the voltage is applied to its terminals. In **Figure 21c**, one of the basic properties of a series motor is seen, which is that in no load condition (here its load is only torque of its mechanical losses, which is about 10% of the rated torque), the field current is strongly suppressed, which results in enormous increasing of the speed.

For this reason, this motor in praxis cannot be in no-load condition and is not recommended to carry out its connection to the load by means of chain, or band, because in the case of a fault, it could be destroyed. In simulation the motor is after the steady condition at the instant $t_1 = 7$ s loaded by its rated torque. In **Figure 21d–f**, mechanical characteristics $n = f(T_e)$ for steady-state conditions are shown, if speed control is carried out by terminal voltage U_q, resistance in the armature circuit R_q (in this case there is also resistance of field circuit), as well as field current i_f (there is a resistance parallelly connected to the field winding).

2.3.3.3 Series dynamo

The approach to the simulations is the same as in previous chapters concerning the generating operations: the constant driving speed is supposed, induced voltage is a source for the whole circuit, and this voltage covers not only the voltage drops in the field and armature windings but also the terminal voltage. The current is only one $i = i_f = i_q$, and the terminal voltage is given also by the load resistance:

$$u = R_L i = \omega L_{df} i - (R_q + R_f) i - (L_q + L_f) \frac{di}{dt}. \tag{119}$$

The magnetizing characteristic, i.e., no load curve $U_i = f(I_f)$, must be measured by a separate excitation.

2.3.3.4 Simulations of a DC series dynamo

Data and parameters of a machine which was simulated in generating operations are in **Table 2**. Dynamo is kept at constant speed; at first in the no load condition, it means terminals are opened, and no current flows in its circuit. A small voltage is possible to measure at its terminals at this condition. This voltage is induced by

$U_{qN} = 180$ V	$R_q = 1\,\Omega$
$I_{qN} = 5$ A	$L_q = 0.005$ mH
$P_N = 925$ W	$R_f = 1\,\Omega$
$n_N = 3000$ min^{-1}	$L_f = 0.015$ H
$M_N = 3$ Nm	$L_{qf} = 0.114$ H
$I_{fN} = 5$ A	$J = 0.003$ kg m^2
$p = 1$	

Table 2.
Nameplate and parameters of simulated series motor.

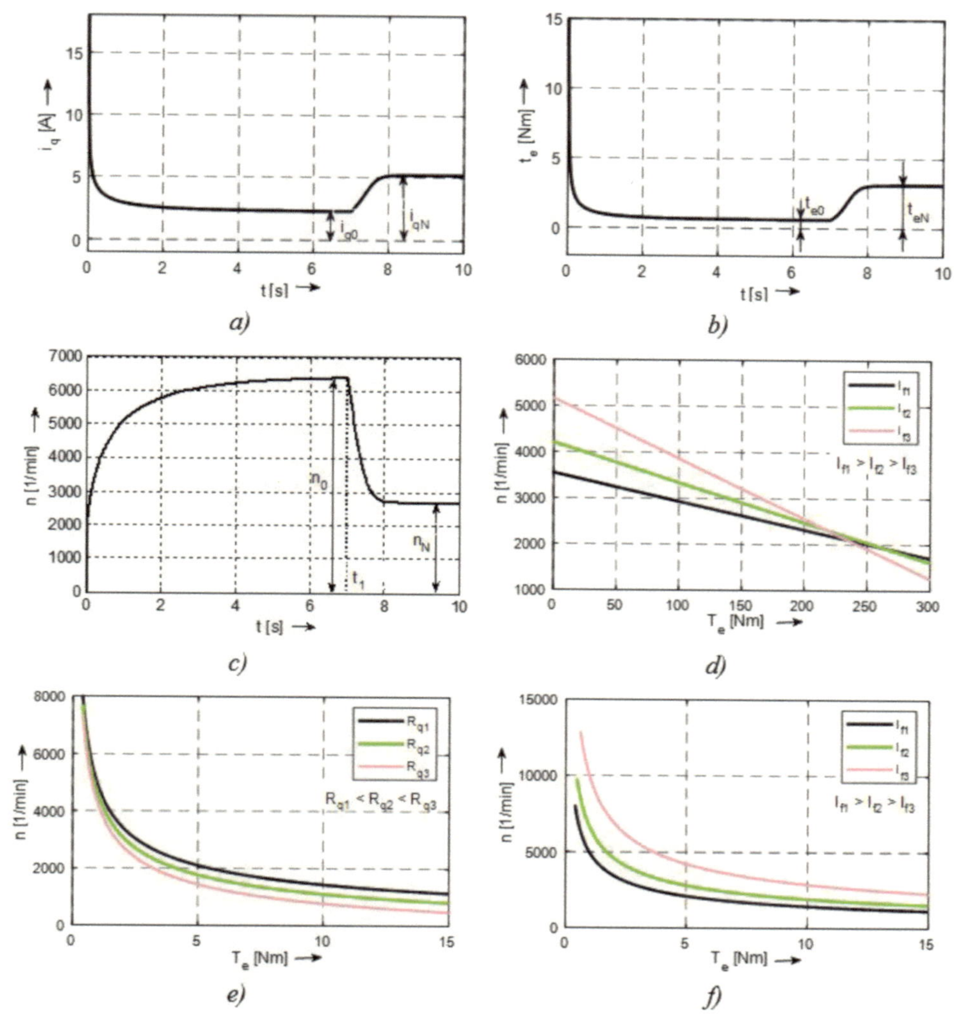

Figure 21.
Simulations of series motor. Time waveforms of (a) armature current and field current, (b) developed electromagnetic torque, and (c) speed and then speed vs. torque in the steady-state conditions for (d) various terminal voltages; (e) various resistances in series with armature circuit, at U_{qN}; and (f) various field currents.

means of remanent magnetic flux (**Figure 23b**, $u_i = f(t)$). For this purpose, it is necessary to measure magnetizing curve at separate excitation $U_i = f(I_f)$. For the investigated machine, this curve is shown in **Figure 22**.

After the load is applied to the terminals at the instant $t_1 = 0.2$ s, the current starts to flow in the circuit, because of the induced voltage (**Figure 23a**), $i_f = i_q = f(t)$, which flows also through the field winding and causes higher excitation of the machine, which results in higher induced voltage. Then the current is increased, which results again in the increasing of the induced voltage, etc. The transients are stabilized after the magnetic circuit is saturated. In this condition the voltage is increased with the increasing of the current, very slowly (**Figure 22**, $U_i = f(I_f)$). Similarly, as induced voltage, also the terminal voltage is increased with the increasing of the current but only till the saturation of the magnetic circuit. Then the terminal voltage can even sink, because the voltage drops on the armature and field resistances can increase quicker than induced voltage. In this simulated case, this did not appear, and the terminal voltage was increased with the increased current (see **Figure 23d** and the curve $U_q = f(I_q)$).

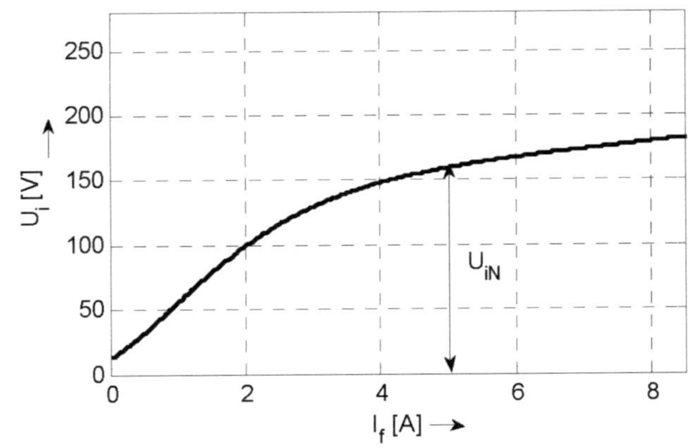

Figure 22.
Measured magnetizing curve for the investigated series machine $U_i = f(I_f)$.

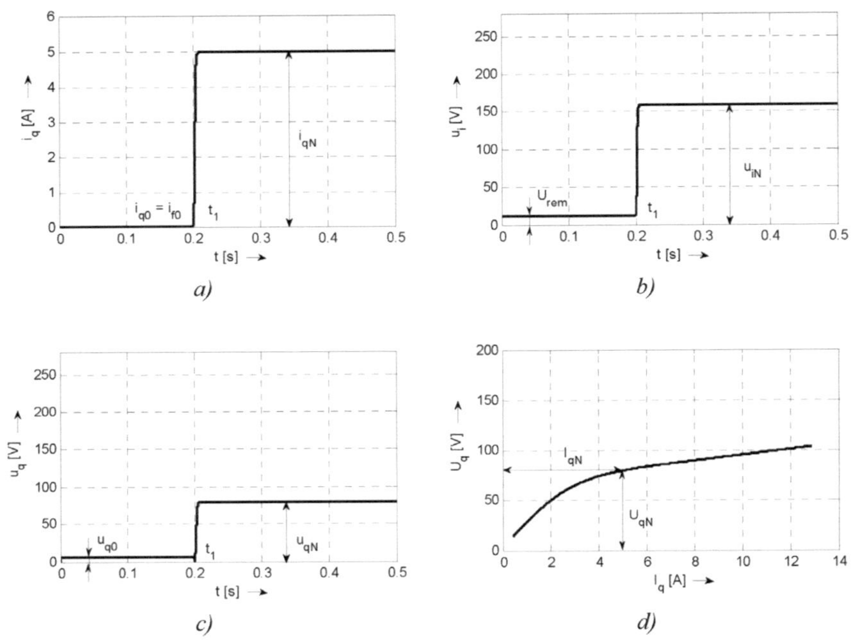

Figure 23.
Simulations of the series dynamo and time waveforms of (a) armature current, (b) induced voltage, (c) terminal voltage, and (d) terminal voltage vs. load current in steady-state conditions at the changing of the load resistance.

2.3.4 Compound machines

As it is known, compound machines are fitted with both series and shunt field windings. Therefore, also simulations of transients and steady-state conditions are made on the basis of combinations of appropriate equations discussed in the previous chapters.

2.3.5 Single-phase commutator series motors

These motors, known as universal motor, can work on DC as well as AC network. Their connection is identical with series DC motors, even though there are some differences in their design. At the simulations, it is necessary to take into account that there are alternating variables of voltage and current; it means that winding's parameters act as impedances, not only resistances.

2.4 Transformation of the three-phase system abc to the system dq0

2.4.1 Introduction

Up to now we have dealt with DC machines, the windings of which are arranged in two perpendicular axes to each other. However, alternating rotating machines obviously have three-phase distributed windings on the stator, which must be transformed into two perpendicular axes, to be able to employ equations derived in the previous chapters.

In history, it can be found that principles of the variable projections into two perpendicular axes were developed for synchronous machine with salient poles.

A different air gap in the axis that acts as field winding and magnetic flux is created and, in the axis perpendicular to that magnetic flux, was linked with a different magnetic permeance of the circuit, which resulted in different reactances of armature reaction and therefore different synchronous reactances. It was shown that this projection into two perpendicular axes and variables can be employed much wider and can be applied for investigation of transients on the basis of the general theory of electrical machines.

On the other side, it is necessary to realize that phase values transformed into dq0 system have gotten into a fictitious system with fictitious parameters, where investigation is easier, but the solution does not show real values. Therefore, an inverse transformation into the abc system must be done to gain real values of voltages, currents, torques, powers, speed, etc. This principle is not unknown in the other investigation of electrical machines. For example, the rotor variables referred to the stator in the case of asynchronous machines mean investigation in a fictitious system, where 29 the calculation and analysis is more simple, but to get real values in the rotor winding a reverse transformation must be done.

Therefore, we will deal with a transformation of the phase variables abc into the fictitious reference k-system dq0 with two perpendicular axes which rotate by angular speed ω_k with regard to the stator system. The axis "0" is perpendicular to the plain given by two axes d, q. As it will be shown, the investigation of the machine properties in this system is simpler because the number of equations is reduced, which is a big advantage. However, to get values of the real variables, it will be necessary to make an inverse transformation, as it will be shown gradually in the next chapters.

A graphical interpretation of the transformation abc into the system dq0 is shown in **Figure 24**. This arrangement is formed according to the original letters given by the papers of R.H. Park and his co-authors (around 1928 and later), e.g., [14], although nowadays it is possible to find various other figures, corresponding to the different position of the axes d, q, and corresponding equations.

According to the original approach, if the three-phase system is symmetrical, the d-axis is shifted from the axis of the a-phase about the angle ϑ_k, and the q-axis is ahead of the d-axis by about 90°; then the components in the d-axis and q-axis are the projections of the phase variables of voltage, linkage magnetic flux, or currents, generally marked as x-variable, into those axes. In the given papers, there are derived equations of the abc into dq0 transformation as well as the equations of the inverse transformation dq0 into abc, because of the investigation of the

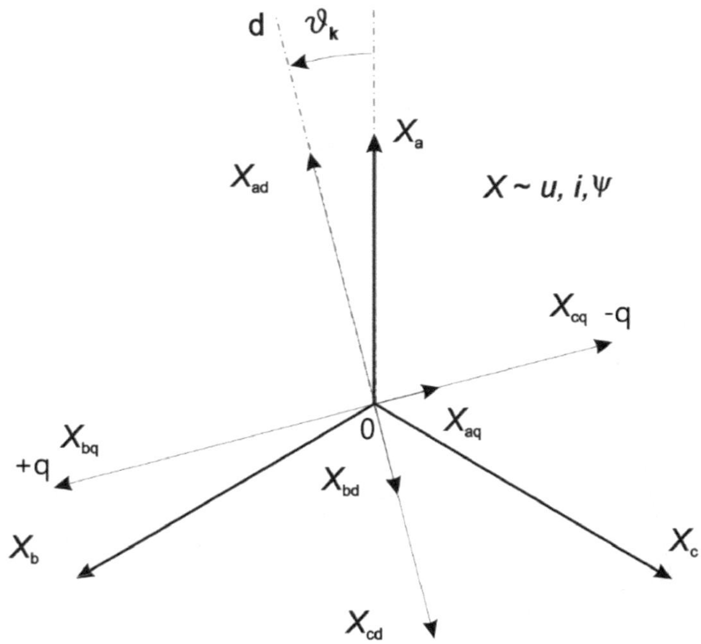

Figure 24.
Graphical interpretation of the three-phase variable transformation abc into the reference k-system dq0, rotating by the speed ω_k.

synchronous reactances of the synchronous machine with salient poles. Also constants of proportions are given. Today this transformation is called "Park's transformation" (see equations given below), even though this name is not given in the original papers. Next equations will be derived, and the constants of proportions k_d, k_q, and k_0 will be employed. Later these constants will be selected according to how the reference system will be positioned, to apply the most profitable solutions. Employment of Park's transformation equations is today very widespread, and they are used for all kinds of electrical machines, frequency convertors, and other three-phase circuits.

2.4.2 Equations of Park's transformations abc into dq0 system

According to **Figure 24**, the d-component of the x-variable is a sum of a-, b-, and c-phase projections:

$$x_d = x_{da} + x_{db} + x_{dc}, \tag{120}$$

where

$$x_{da} = x_a \cos \vartheta_k, \tag{121}$$

$$x_{db} = x_b \cos(\vartheta_k - 120°), \tag{122}$$

$$x_{dc} = x_c \cos(\vartheta_k + 120°). \tag{123}$$

Also, projections into the q-axis are made in a similar way. It is seen that the projections to the q-axis are expressed by sinusoidal function of the phase variable with a negative sign, at the given +q-axis (see Eq. (125)).

The zero component is a sum of the instantaneous values of the phase variables. If the three-phase system is symmetrical, the sum of the instantaneous values is zero; therefore also the zero component is zero (see Eq. (126)). The zero component can be visualized in such a way that the three-phase variable projection is made in the 0-axis perpendicular to the plain created by the d-axis and q-axis, whereby the 0-axis is conducted through the point 0.

Then the equation system for the Park transformation from the abc to the dq0 system is created by Eqs. (124)–(126). To generalize the expressions, proportional constants k_d, k_q, and k_0 are employed:

$$x_d = k_d\left(x_a \cos \vartheta_k + x_b \cos\left(\vartheta_k - \frac{2\pi}{3}\right) + x_c \cos\left(\vartheta_k + \frac{2\pi}{3}\right)\right), \tag{124}$$

$$x_q = -k_q\left(x_a \sin \vartheta_k + x_b \sin\left(\vartheta_k - \frac{2\pi}{3}\right) + x_c \sin\left(\vartheta_k + \frac{2\pi}{3}\right)\right), \tag{125}$$

$$x_0 = k_0(x_a + x_b + x_c). \tag{126}$$

It is true that R.H. Park does not mention such constants in the original paper, because he solved synchronous machine, which will be explained later (Sections 8, 10, and 16). For the purposes of this textbook, it is suitable to start as general as possible and gradually adapt the equations to the individual kinds of electrical machines to get a solution as advantageous as possible. Therefore, the constants can be whichever except zero, though of such, that the equation determinant is not zero (see Eq. (127)). Then the inverse transformation will be possible to do and to find the real phase variables.

The determinant of the system is as follows:

$$\begin{vmatrix} k_d \cos \vartheta_k & k_d \cos\left(\vartheta_k - \frac{2\pi}{3}\right) & k_d \cos\left(\vartheta_k + \frac{2\pi}{3}\right) \\ -k_q \sin \vartheta_k & -k_q \sin\left(\vartheta_k - \frac{2\pi}{3}\right) & -k_q \sin\left(\vartheta_k + \frac{2\pi}{3}\right) \\ k_0 & k_0 & k_0 \end{vmatrix} = k_d k_q k_0 \frac{3\sqrt{3}}{2} \cos\left(\vartheta_k - \frac{2\pi}{3}\right). \tag{127}$$

2.4.3 Equations for the m-phase system transformation

Equations for the three-phase system transformation can be spread to the m-phase system. Now the phases will be marked by 1, 2, 3, etc., to be able to express the mth phase and to see how the argument of the functions is created:

$$x_d = k_d\left(x_1 \cos \vartheta_k + x_2 \cos\left(\vartheta_k - \frac{2\pi}{m}\right) + x_3 \cos\left(\vartheta_k - \frac{4\pi}{m}\right) + \dots + x_m \cos\left(\vartheta_k - \frac{2(m-1)\pi}{m}\right)\right). \tag{128}$$

Similarly, equations for the q- and 0-components are written. If a proportional constant 2/3 will be used for the three-phase system, then the corresponding constant for the m-phase system is $2/m$ [2].

$$x_d = \frac{2}{m}\left(x_1 \cos \vartheta_k + x_2 \cos\left(\vartheta_k - \frac{2\pi}{m}\right) + x_3 \cos\left(\vartheta_k - \frac{4\pi}{m}\right) + \dots + x_m \cos\left(\vartheta_k - \frac{2(m-1)\pi}{m}\right)\right). \tag{129}$$

2.5 Inverse transformation from dq0 to the abc system

Equations for the inverse transformation are derived from the previous equations. Equation (124) is multiplied by expression $\cos\vartheta_k/k_d$ and added to Eq. (125), which was multiplied by the expression $-\sin\vartheta_k/k_q$. After the modification it is:

$$\frac{x_d\cos\vartheta_k}{k_d}-\frac{x_q\sin\vartheta_k}{k_q}=x_a-\frac{1}{2}x_b-\frac{1}{2}x_c=x_a-\frac{1}{2}(x_b+x_c), \tag{130}$$

and from the third Eq. (126), the following is derived:

$$\frac{x_0}{k_0}=x_a+x_b+x_c\rightarrow\frac{x_0}{k_0}-x_a=x_b+x_c, \tag{131}$$

which is necessary to introduce to Eq. (130):

$$x_a-\frac{1}{2}\left(\frac{x_0}{k_0}-x_a\right)=\frac{3}{2}x_a-\frac{1}{2}\left(\frac{x_0}{k_0}\right)=\frac{x_d\cos\vartheta_k}{k_d}-\frac{x_q\sin\vartheta_k}{k_q}. \tag{132}$$

In this way, the equation for the inverse transformation of the a-phase variable is gained:

$$x_a=\frac{2}{3}\frac{1}{k_d}x_d\cos\vartheta_k-\frac{2}{3}\frac{1}{k_q}x_q\sin\vartheta_k+\frac{1}{3}\frac{1}{k_0}x_0. \tag{133}$$

In a similar way, equations for the inverse transformation and also for b-phase and c-phase are derived:

$$x_b=\frac{2}{3}\frac{1}{k_d}x_d\cos\left(\vartheta_k-\frac{2\pi}{3}\right)-\frac{2}{3}\frac{1}{k_q}x_q\sin\left(\vartheta_k-\frac{2\pi}{3}\right)+\frac{1}{3}\frac{1}{k_0}x_0, \tag{134}$$

$$x_c=\frac{2}{3}\frac{1}{k_d}x_d\cos\left(\vartheta_k+\frac{2\pi}{3}\right)-\frac{2}{3}\frac{1}{k_q}x_q\sin\left(\vartheta_k+\frac{2\pi}{3}\right)+\frac{1}{3}\frac{1}{k_0}x_0. \tag{135}$$

Equations (133)–(135) create a system for the inverse transformation from dq0 to the abc system. These equations will be employed, e.g., for calculation of the real currents in the phase windings, if the currents in the dq0 system are known.

2.6 Equations of the linear transformation made by means of the space vectors of the voltage and currents

A space vector is a formally introduced symbol, which is illustrated in a complex plain in such a way that its position determines space position of the positive maximum of the total magnetic flux or magnetic flux density.

This definition is very important because as we know from the theory of electromagnetic field, neither current nor voltage is the vector. After the definition of the space vectors, it is possible to work with the currents and voltages, linked by Ohm's law through impedance, but to image that it is a vector of the air gap magnetic flux density, which is by these currents and voltages created, which is very profitable. Therefore to distinguish a term "vector" as a variable which has a value and a direction, here the term "space vector" is used. The whole name "space vector" should be expressed and should not be shortened to "vector" because it can cause a misunderstanding, mainly between the people who do not work with investigation of transients.

To express that all three phases to which terminal voltages u_a, u_b, and u_c are applied and contribute to the creation of the air gap magnetic field and magnetic flux density, it is possible to use the equation of the voltage space vector. In the complex plain, it will represent the value and position of air gap magnetic flux density magnitude:

$$\bar{u}_s = k_s(u_a + \bar{a}u_b + \bar{a}^2 u_c), \tag{136}$$

where unit phasors $\bar{a}$ mean a shift of the voltage phasor about 120° (note: phasor shows time shifting of variables):

$$\bar{a} = e^{j\frac{2\pi}{3}} = \cos\frac{2\pi}{3} + j\sin\frac{2\pi}{3}, \tag{137}$$

$$\bar{a}^2 = e^{j\frac{4\pi}{3}} = \cos\frac{4\pi}{3} + j\sin\frac{4\pi}{3} = e^{-j\frac{2\pi}{3}} = \cos\frac{2\pi}{3} - j\sin\frac{2\pi}{3}. \tag{138}$$

The subscript "s" means that it is a stator variable. Also, a proportional constant is marked with this subscript. In **Figure 25**, a complex plain with the stator axis is graphically illustrated, which is now identical with the axis of the a-phase winding. Then there is a rotor axis, which is shifted from the stator axis about the ϑ_r angle, and the axis of the k-reference frame, which is shifted from the stator axis about an arbitrary ϑ_k angle. Between the rotor axis and axis of the k-reference frame, there is an angle $(\vartheta_k - \vartheta_r)$. The axis of the k-reference frame is identical with its real component in the d-axis, and this system rotates by the angular speed ω_k in the marked direction. The space vector of the stator voltage can be written as a sum of its real and imaginary components:

$$\bar{u}_s = u_d \pm ju_q. \tag{139}$$

2.6.1 Stator variable transformation

The transformation of the stator variables into the k-reference frame $(k, +jk)$ means to multiply stator variables by the expression $e^{-j\vartheta_k}$; it means the k-axis must be shifted back about the angle $-\vartheta_k$, to identify it with the stator axis:

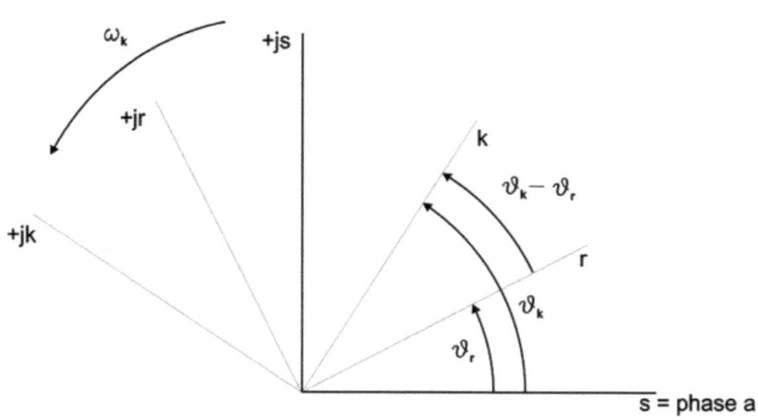

Figure 25.
Graphical illustration of the complex plain with the stator axis, rotor axis, and axis of the k-reference frame.

$$\bar{u}_\mathrm{s}^\mathrm{k} = k_\mathrm{s}(u_\mathrm{a} + \bar{a}u_\mathrm{b} + \bar{a}^2 u_\mathrm{c})e^{-j\vartheta_\mathrm{k}} = k_\mathrm{s}\left(u_\mathrm{a}e^{-j\vartheta_\mathrm{k}} + u_\mathrm{b}e^{-j\left(\vartheta_\mathrm{k}-\frac{2\pi}{3}\right)} + u_\mathrm{c}e^{-j\left(\vartheta_\mathrm{k}+\frac{2\pi}{3}\right)}\right). \quad (140)$$

If this equation is split by means of the goniometrical functions into the real components, i.e., with cos-members, and into the imaginary components with sin-members,

$$e^{-j\vartheta_\mathrm{k}} = \cos\vartheta_\mathrm{k} - j\sin\vartheta_\mathrm{k}, \quad (141)$$

then in a coincidence with Eqs. (139) and (141), the next two equations are gained:

$$u_\mathrm{d} = k_\mathrm{s}\left(u_\mathrm{a}\cos\vartheta_\mathrm{k} + u_\mathrm{b}\cos\left(\vartheta_\mathrm{k} - \frac{2\pi}{3}\right) + u_\mathrm{c}\cos\left(\vartheta_\mathrm{k} + \frac{2\pi}{3}\right)\right), \quad (142)$$

$$u_\mathrm{q} = -k_\mathrm{s}\left(u_\mathrm{a}\sin\vartheta_\mathrm{k} + u_\mathrm{b}\sin\left(\vartheta_\mathrm{k} - \frac{2\pi}{3}\right) + u_\mathrm{c}\sin\left(\vartheta_\mathrm{k} + \frac{2\pi}{3}\right)\right). \quad (143)$$

As it is seen, these equations are identical with Eqs. (124) and (125), which were derived in general for three-phase circuits. In those equations separately marked constants in each axis were introduced, but here it is justified that it is enough to employ only one constant for both axes:

$$k_\mathrm{d} = k_\mathrm{q} = k_\mathrm{s}. \quad (144)$$

The third equation for the zero component, which is needed for investigation of the asymmetrical systems, can be added:

$$u_0 = k_0(u_\mathrm{a} + u_\mathrm{b} + u_\mathrm{c}). \quad (145)$$

2.6.2 Rotor variable transformation

The same approach is used for the rotor variables with the subscript "r." To distinguish them from the stator variables, the subscripts DQ0 will be used for the k-axis and ABC for the phase variables. The rotor variables must be multiplied by the expression $e^{-j(\vartheta_\mathrm{k}-\vartheta_\mathrm{r})}$ if rotor variables are transformed into the k-axis. It means the k-axis must be shifted back about the angle $(\vartheta_\mathrm{k} - \vartheta_\mathrm{r})$, to identify it with the rotor axis. In general, space vector of the rotor voltage transformed into k-system can be written as follows:

$$\begin{aligned}\bar{u}_\mathrm{r}^\mathrm{k} &= k_\mathrm{r}(u_\mathrm{A} + \bar{a}u_\mathrm{B} + \bar{a}^2 u_\mathrm{C})e^{-j(\vartheta_\mathrm{k}-\vartheta_\mathrm{r})} \\ &= k_\mathrm{r}\left(u_\mathrm{A}e^{-j(\vartheta_\mathrm{k}-\vartheta_\mathrm{r})} + u_\mathrm{B}e^{-j\left(\vartheta_\mathrm{k}-\vartheta_\mathrm{r}-\frac{2\pi}{3}\right)} + u_\mathrm{C}e^{-j\left(\vartheta_\mathrm{k}-\vartheta_\mathrm{r}+\frac{2\pi}{3}\right)}\right).\end{aligned} \quad (146)$$

The variables of the k-system expressed in two perpendicular d-axis and q-axis are:

$$\bar{u}_\mathrm{r}^\mathrm{k} = u_\mathrm{D} \pm ju_\mathrm{Q}; \quad (147)$$

then Eq. (146) can be itemized into two equations:

$$u_\mathrm{D} = k_\mathrm{r}\left(u_\mathrm{A}\cos(\vartheta_\mathrm{k} - \vartheta_\mathrm{r}) + u_\mathrm{B}\cos\left(\vartheta_\mathrm{k} - \vartheta_\mathrm{r} - \frac{2\pi}{3}\right) + u_\mathrm{C}\cos\left(\vartheta_\mathrm{k} - \vartheta_\mathrm{r} + \frac{2\pi}{3}\right)\right), \quad (148)$$

$$u_{\mathrm{Q}} = -k_{\mathrm{r}}\left(u_{\mathrm{A}}\sin\left(\vartheta_{\mathrm{k}} - \vartheta_{\mathrm{r}}\right) + u_{\mathrm{B}}\sin\left(\vartheta_{\mathrm{k}} - \vartheta_{\mathrm{r}} - \frac{2\pi}{3}\right) + u_{\mathrm{C}}\sin\left(\vartheta_{\mathrm{k}} - \vartheta_{\mathrm{r}} + \frac{2\pi}{3}\right)\right) \tag{149}$$

and for the zero component:

$$u_{\mathrm{O}} = k_{\mathrm{O}}(u_{\mathrm{A}} + u_{\mathrm{B}} + u_{\mathrm{C}}). \tag{150}$$

Here is a system of equations for rotor variables of the three-phase system ABC transformed into the k-system DQ0. They differ from the stator variables by the angle $(\vartheta_{\mathrm{k}} - \vartheta_{\mathrm{r}})$ instead of ϑ_{k}.

2.7 Voltage equations of three-phase machines and their windings

The same equations as for the terminal voltage of the universal machine (63) can be written also for the terminal voltage of the three-phase machines. For example, the stator windings, where the phases are marked with a, b, and c equations of the terminal voltage, are in the next form:

$$u_{\mathrm{a}} = R_{\mathrm{a}} i_{\mathrm{a}} + \frac{\mathrm{d}\psi_{\mathrm{a}}}{\mathrm{d}t}, \tag{151}$$

$$u_{\mathrm{b}} = R_{\mathrm{b}} i_{\mathrm{b}} + \frac{\mathrm{d}\psi_{\mathrm{b}}}{\mathrm{d}t}, \tag{152}$$

$$u_{\mathrm{c}} = R_{\mathrm{c}} i_{\mathrm{c}} + \frac{\mathrm{d}\psi_{c}}{\mathrm{d}t}. \tag{153}$$

In the next only a-phase will be investigated. In Eq. (151), the variables u_{a}, i_{a}, and ψ_{a} will be introduced; after they are adapted according to Eq. (133), it means on the basis of inverse transformation (Section 5). For example, for the ψ_{a} it is expressed in the form:

$$\psi_{\mathrm{a}} = \frac{2}{3}\frac{1}{k_{\mathrm{d}}}\psi_{\mathrm{d}}(\cos\vartheta_{\mathrm{k}}) - \frac{2}{3}\frac{1}{k_{\mathrm{q}}}\psi_{\mathrm{q}}(\sin\vartheta_{\mathrm{k}}) + \frac{1}{3}\frac{1}{k_{0}}\psi_{0}. \tag{154}$$

Now a derivation by time $\mathrm{d}\psi_{\mathrm{a}}/\mathrm{d}t$ is made:

$$\begin{aligned}\frac{\mathrm{d}\psi_{\mathrm{a}}}{\mathrm{d}t} = {} & \frac{2}{3}\frac{1}{k_{\mathrm{d}}}\frac{\mathrm{d}\psi_{\mathrm{d}}}{\mathrm{d}t}\cos\vartheta_{\mathrm{k}} - \frac{2}{3}\frac{1}{k_{\mathrm{d}}}\omega_{\mathrm{k}}\psi_{\mathrm{d}}\sin\vartheta_{\mathrm{k}} - \frac{2}{3}\frac{1}{k_{\mathrm{q}}}\frac{\mathrm{d}\psi_{\mathrm{q}}}{\mathrm{d}t}\sin\vartheta_{\mathrm{k}} - \frac{2}{3}\frac{1}{k_{\mathrm{q}}}\omega_{\mathrm{k}}\psi_{\mathrm{q}}\cos\vartheta_{\mathrm{k}} \\ & + \frac{1}{3}\frac{1}{k_{0}}\frac{\mathrm{d}\psi_{0}}{\mathrm{d}t}\end{aligned} \tag{155}$$

From Eq. (151), $\mathrm{d}\psi_{\mathrm{a}}/\mathrm{d}t$ is selected:

$$\frac{\mathrm{d}\psi_{\mathrm{a}}}{\mathrm{d}t} = u_{\mathrm{a}} - R_{\mathrm{s}} i_{\mathrm{a}}. \tag{156}$$

It was supposed that all three phases are identical and their resistances are equal:

$$R_{\mathrm{a}} = R_{\mathrm{b}} = R_{\mathrm{c}} = R_{\mathrm{s}}. \tag{157}$$

The expressions from the inverse transformation are introduced also for u_a and i_a, in Eq. (156):

$$\frac{d\psi_a}{dt} = \frac{2}{3}\frac{1}{k_d}u_d \cos\vartheta_k - \frac{2}{3}\frac{1}{k_q}u_q \sin\vartheta_k + \frac{1}{3}\frac{1}{k_0}u_0 - R_s\left(\frac{2}{3}\frac{1}{k_d}i_d \cos\vartheta_k - \frac{2}{3}\frac{1}{k_q}i_q \sin\vartheta_k + \frac{1}{3}\frac{1}{k_0}i_0\right) \quad (158)$$

The left sides of Eqs. (155) and (158) are equal; therefore, also right sides will be equal. Now the members with the same goniometrical functions and members without goniometrical functions will be selected and put equal, e.g., members at *cos* ϑ_k yield:

$$\frac{2}{3}\frac{1}{k_d}\frac{d\psi_d}{dt} - \frac{2}{3}\frac{1}{k_q}\omega_k\psi_q = \frac{2}{3}\frac{1}{k_d}u_d - R_s\frac{2}{3}\frac{1}{k_d}i_d. \quad (159)$$

The equation for u_d is gained after modification, and in a very similar way, also the two other equations are obtained:

$$u_d = R_s i_d + \frac{d\psi_d}{dt} - \omega_k\psi_q, \quad (160)$$

$$u_q = R_s i_q + \frac{d\psi_q}{dt} + \omega_k\psi_d, \quad (161)$$

$$u_0 = R_s i_0 + \frac{d\psi_0}{dt}. \quad (162)$$

Equations (160)–(162) are the voltage equations for the stator windings of the three-phase machines, such as asynchronous motors in k-reference frame, rotating by the angular speed ω_k with the dq0-axis. As it can be seen, they are the same equations as the voltage equations in Section 2.1, which were derived for universal arrangement of the electrical machine. Here general validity of the equations is seen: if windings of any machine are arranged or are transformed to the arrangement with two perpendicular axes to each other, the same equations are valid. Of course, the parameters, mainly inductances, of the machine are different, and it is necessary to know how to get them.

2.8 Three-phase power and torque in the system dq0

2.8.1 Three-phase power in the system dq0

The instantaneous value of the input power in a three-phase system is a sum of instantaneous values of power in each phase (see also Section 2.2):

$$p_{in} = u_a i_a + u_b i_b + u_c i_c. \quad (163)$$

Instantaneous values of u_a and i_a will be introduced into this equation. These were derived in the inverse transformation chapter. They are Eqs. (133)–(135). It means at first u is introduced for x, and it must be multiplied by expression, where i was introduced, and then the further phases in the same way are adapted. At the end all expressions are summed:

$$p_{\text{in}} = \left[\frac{2}{3}\frac{1}{k_{\text{d}}}u_{\text{d}}\cos\vartheta_{\text{k}} - \frac{2}{3}\frac{1}{k_{\text{q}}}u_{\text{q}}\sin\vartheta_{\text{k}} + \frac{1}{3}\frac{1}{k_0}u_0\right]\left[\frac{2}{3}\frac{1}{k_{\text{d}}}i_{\text{d}}\cos\vartheta_{\text{k}} - \frac{2}{3}\frac{1}{k_{\text{q}}}i_{\text{q}}\sin\vartheta_{\text{k}} + \frac{1}{3}\frac{1}{k_0}i_0\right]$$
$$+\left[\frac{2}{3}\frac{1}{k_{\text{d}}}u_{\text{d}}\cos\left(\vartheta_{\text{k}} - \frac{2\pi}{3}\right) - \frac{2}{3}\frac{1}{k_{\text{q}}}u_{\text{q}}\sin\left(\vartheta_{\text{k}} - \frac{2\pi}{3}\right) + \frac{1}{3}\frac{1}{k_0}u_0\right]$$
$$\cdot\left[\frac{2}{3}\frac{1}{k_{\text{d}}}i_{\text{d}}\cos\left(\vartheta_{\text{k}} - \frac{2\pi}{3}\right) - \frac{2}{3}\frac{1}{k_{\text{q}}}i_{\text{q}}\sin\left(\vartheta_{\text{k}} - \frac{2\pi}{3}\right) + \frac{1}{3}\frac{1}{k_0}i_0\right]$$
$$+\left[\frac{2}{3}\frac{1}{k_{\text{d}}}u_{\text{d}}\cos\left(\vartheta_{\text{k}} + \frac{2\pi}{3}\right) - \frac{2}{3}\frac{1}{k_{\text{q}}}u_{\text{q}}\sin\left(\vartheta_{\text{k}} + \frac{2\pi}{3}\right) + \frac{1}{3}\frac{1}{k_0}u_0\right]$$
$$\cdot\left[\frac{2}{3}\frac{1}{k_{\text{d}}}i_{\text{d}}\cos\left(\vartheta_{\text{k}} + \frac{2\pi}{3}\right) - \frac{2}{3}\frac{1}{k_{\text{q}}}i_{\text{q}}\sin\left(\vartheta_{\text{k}} + \frac{2\pi}{3}\right) + \frac{1}{3}\frac{1}{k_0}i_0\right] \tag{164}$$

Now it is necessary to multiply all members with each other, including the goniometrical functions, and after a modification the result is:

$$p_{\text{in}} = \left[\frac{2}{3}\frac{1}{k_{\text{d}}^2}u_{\text{d}}i_{\text{d}} + \frac{2}{3}\frac{1}{k_{\text{q}}^2}u_{\text{q}}i_{\text{q}} + \frac{1}{3}\frac{1}{k_0^2}u_0i_0\right]. \tag{165}$$

Variables u_{d}, u_{q}, and u_0 are given by Eqs. (160) to (162), which were introduced above, and after a modification, the result is:

$$p_{\text{in}} = \left[\frac{2}{3}\frac{1}{k_{\text{d}}^2}\left(R_{\text{s}}i_{\text{d}}^2 + i_{\text{d}}\frac{\text{d}\psi_{\text{d}}}{\text{d}t} - \omega_{\text{k}}\psi_{\text{q}}i_{\text{d}}\right) + \frac{2}{3}\frac{1}{k_{\text{q}}^2}\left(R_{\text{s}}i_{\text{q}}^2 + i_{\text{q}}\frac{\text{d}\psi_{\text{q}}}{\text{d}t} + \omega_{\text{k}}\psi_{\text{d}}i_{\text{q}}\right) + \frac{1}{3}\frac{1}{k_0^2}\left(R_{\text{s}}i_0^2 + i_0\frac{\text{d}\psi_0}{\text{d}t}\right)\right] \tag{166}$$

If an analysis in greater details is made, it is seen that an input power on the left side must be in equilibrium with the right side. It is supposed to be motoring operation. Therefore the input power applied to the terminals of the three-phase motor is distributed between the Joule's resistance loss (ΣRi^2), time varying of the field energy stored in the investigated circuit $\Sigma i\frac{\text{d}\psi}{\text{d}t}$, and the rest of the members' mean conversion of electrical to mechanical energy and eventually to mechanical output power. If the resistance loss and power of magnetic field are subtracted from the input power on the terminals, the result is an air gap electromagnetic power, which is given by the difference of two rotating voltages in both axes:

$$p_{\text{e}} = \frac{2}{3}\left(\frac{1}{k_{\text{q}}^2}\omega_{\text{k}}\psi_{\text{d}}i_{\text{q}} - \frac{1}{k_{\text{q}}^2}\omega_{\text{k}}\psi_{\text{q}}i_{\text{d}}\right). \tag{167}$$

Here it is seen that it is advantageous to choose the same proportional constants: $k_{\text{d}} = k_{\text{q}}$, to be able to set out it in front of the brackets, together with the angular speed:

$$p_{\text{e}} = \frac{2}{3}\frac{1}{k_{\text{q}}^2}\omega_{\text{k}}\left(\psi_{\text{d}}i_{\text{q}} - \psi_{\text{q}}i_{\text{d}}\right) \tag{168}$$

and, eventually,

$$p_{\text{e}} = \frac{2}{3}\frac{1}{k_{\text{d}}k_{\text{q}}}\omega_{\text{k}}\left(\psi_{\text{d}}i_{\text{q}} - \psi_{\text{q}}i_{\text{d}}\right). \tag{169}$$

This is the base expression for the power, which is converted from an electrical to a mechanical form in the motor or from a mechanical to an electrical form in the case of the generator. Next an expression for the electromagnetic torque is derived.

2.8.2 Electromagnetic torque of the three-phase machines in the dq0 system

As it is known, an air gap power can be expressed by the product of the developed electromagnetic torque and a mechanical angular speed, now in the k-system:

$$p_e = t_e \Omega_k \tag{170}$$

or by means of electrical angular speed:

$$p_e = t_e \frac{\omega_k}{p} \tag{171}$$

where p is the number of pole pairs. An instantaneous value of the developed electromagnetic torque is valid:

$$t_e = \frac{p}{\omega_k} p_e = \frac{p}{\omega_k} \frac{2}{3} \frac{1}{k_d k_q} \omega_k \left(\psi_d i_q - \psi_q i_d \right) \tag{172}$$

and after a reduction the torque is:

$$t_e = p \frac{2}{3} \frac{1}{k_d k_q} \left(\psi_d i_q - \psi_q i_d \right). \tag{173}$$

This is the base expression for an instantaneous value of developed electromagnetic torque of a three-phase machine. It is seen that its concrete form will be modified according to the chosen proportional constants. The most advantageous choice seems to be the next two possibilities:

(1) $k_d = k_q = \frac{2}{3}, k_0 = \frac{1}{3}$.

Then:

$$t_e = p \frac{2}{3} \frac{1}{\frac{2}{3}\frac{2}{3}} \left(\psi_d i_q - \psi_q i_d \right) = p \frac{3}{2} \left(\psi_d i_q - \psi_q i_d \right) \tag{174}$$

(2) $k_d = k_q = \sqrt{\frac{2}{3}}, k_0 = \sqrt{\frac{1}{3}}$.

Then:

$$t_e = p \frac{2}{3} \frac{1}{\sqrt{\frac{2}{3}}\sqrt{\frac{2}{3}}} \left(\psi_d i_q - \psi_q i_d \right) = p \left(\psi_d i_q - \psi_q i_d \right). \tag{175}$$

It will be shown later that the first choice is more advantageous for asynchronous machines and the second one for synchronous machine.

A developed electromagnetic torque in the rotating electrical machines directly relates with equilibrium of the torques acting on the shaft. During the transients in motoring operation, i.e., when the speed is changing, developed electromagnetic torque t_e covers not only load torque t_L, including loss torque, but also load created by the moment of inertia of rotating mass $J \frac{d\Omega}{dt}$. Therefore, it is possible to write:

$$t_e = J\frac{d\Omega}{dt} + t_L. \tag{176}$$

Unknown variables in motoring operation are obviously currents and speed, which can be eliminated from Eqs. (176) and (175). The mechanical angular speed is valid:

$$\frac{d\Omega}{dt} = \frac{1}{J}(t_e - t_L) \tag{177}$$

and electrical angular speed is:

$$\frac{d\omega}{dt} = \frac{p}{J}(t_e - t_L). \tag{178}$$

The final expression for the time changing of the speed will be gotten, if for t_e Eqs. (174) and (175) according to the choice of the constants k_d and k_q are introduced:

(1) $k_d = k_q = \frac{2}{3}, k_0 = \frac{1}{3}$

Then

$$t_e = p\frac{3}{2}\left(\psi_d i_q - \psi_q i_d\right) \tag{179}$$

$$\frac{d\omega}{dt} = \frac{p}{J}\left(p\frac{3}{2}\left(\psi_d i_q - \psi_q i_d\right) - t_L\right) \tag{180}$$

(2) $k_d = k_q = \sqrt{\frac{2}{3}}, k_0 = \sqrt{\frac{1}{3}}.$

Then

$$t_e = p\left(\psi_d i_q - \psi_q i_d\right) \tag{181}$$

$$\frac{d\omega}{dt} = \frac{p}{J}\left(p\left(\psi_d i_q - \psi_q i_d\right) - t_L\right). \tag{182}$$

If there is a steady-state condition, $\frac{d\omega}{dt} = 0$, and electromagnetic and load torque are in balance:

$$t_e = t_L. \tag{183}$$

2.8.3 Power invariance principle

The expression for the three-phase power in dq0 system is:

$$p_{in} = \left[\frac{2}{3}\frac{1}{k_d^2}u_d i_d + \frac{2}{3}\frac{1}{k_q^2}u_q i_q + \frac{1}{3}\frac{1}{k_0^2}u_0 i_0\right] \tag{184}$$

which was derived from the original expression for the three-phase power in abc system:

$$p_{in} = [u_a i_a + u_b i_b + u_c i_c]. \tag{185}$$

The expression can be modified by means of the constants k_d and k_q:

1. If $k_\mathrm{d} = k_\mathrm{q} = \frac{2}{3}, k_0 = \frac{1}{3}$,

then

$$p_\mathrm{in} = \frac{3}{2}u_\mathrm{d}i_\mathrm{d} + \frac{3}{2}u_\mathrm{q}i_\mathrm{q} + 3u_0 i_0, \tag{186}$$

in which the principle of power invariance is not fulfilled, because the members in dq0 axes are figures, although it was derived from Eq. (163), where no figures were employed.

2. If $k_\mathrm{d} = k_\mathrm{q} = \sqrt{\frac{2}{3}}, k_0 = \sqrt{\frac{1}{3}}$,

then

$$p_\mathrm{in} = u_\mathrm{d}i_\mathrm{d} + u_\mathrm{q}\mathrm{i}_\mathrm{q} + u_0 i_0, \tag{187}$$

in which the principle of power invariance is fulfilled.

2.9 Properties of the transformed sinusoidal variables

In Section 7, the three-phase system abc into the dq0 system was transformed, and expressions for u_d, u_q, and u_0 variables were derived. Now it is necessary to know what must be introduced for u_d, u_q, and u_0, if variables u_a, u_b, and u_c are sinusoidal variables (or also cosinusoidal variables can be taken). It means sinusoidal variables will be transformed from abc to dq0 followed by the rules given in Section 4.2.

Consider the voltage symmetrical three-phase system:

$$u_\mathrm{a} = U_\mathrm{max} \sin \omega_\mathrm{s} t, \tag{188}$$

$$u_\mathrm{b} = U_\mathrm{max} \sin \left(\omega_\mathrm{s} t - \frac{2\pi}{3} \right), \tag{189}$$

$$u_\mathrm{c} = U_\mathrm{max} \sin \left(\omega_\mathrm{s} t + \frac{2\pi}{3} \right), \tag{190}$$

where ω_s is the angular frequency of the stator voltages (and currents). In **Figure 25**, the relationship between the stator, rotor, and k-system is seen. As it was proclaimed, the stator axis is identified with the axis of the stator winding of phase a, the rotor axis is identified with the axis of the rotor winding of phase A, and this axis is shifted from the stator axis about angle ϑ_r. The axis of the reference k-system, to which the stator variables, now voltages, will be transformed, is shifted from the stator axis about the angle ϑ_k and from the rotor axis about the angle $(\vartheta_\mathrm{k} - \vartheta_\mathrm{r})$. The angle of the k-system ϑ_k is during the transients expressed as integral of its angular speed with the initial position $\vartheta_{\mathrm{k}0}$:

$$\vartheta_\mathrm{k} = \int_0^t \omega_\mathrm{k} \mathrm{d}t + \vartheta_{\mathrm{k}0}. \tag{191}$$

Equations for transformation (124) till (126), derived in Section 4 for the variable x, now are applied for the voltage:

$$u_\mathrm{d} = k_\mathrm{d}\left(u_\mathrm{a} \cos \vartheta_\mathrm{k} + u_\mathrm{b} \cos \left(\vartheta_\mathrm{k} - \frac{2\pi}{3} \right) + u_\mathrm{c} \cos \left(\vartheta_\mathrm{k} + \frac{2\pi}{3} \right) \right), \tag{192}$$

$$u_q = -k_q\left(u_a \sin\vartheta_k + u_b \sin\left(\vartheta_k - \frac{2\pi}{3}\right) + u_c \sin\left(\vartheta_k + \frac{2\pi}{3}\right)\right), \tag{193}$$

$$u_0 = k_0(u_a + u_b + u_c), \tag{194}$$

For the phase voltages, expressions from Eqs. (188)–(190) are introduced. At first, adjust expression for the voltage in the d-axis is as follows:

$$u_d = k_d U_{max}\left(\sin\omega_s t\cos\vartheta_k + \sin\left(\omega_s t - \frac{2\pi}{3}\right)\cos\left(\vartheta_k - \frac{2\pi}{3}\right) + \sin\left(\omega_s t + \frac{2\pi}{3}\right)\cos\left(\vartheta_k + \frac{2\pi}{3}\right)\right). \tag{195}$$

After the modification of the goniometrical functions and summarization of the appropriate members, in the final phase, it can be adjusted as follows:

$$\begin{aligned} u_d &= k_d U_{max}\left(\sin\omega_s t\cos\vartheta_k + \frac{1}{2}\sin\omega_s t\cos\vartheta_k - \frac{3}{2}\cos\omega_s t\sin\vartheta_k\right) \\ &= k_d U_{max}\left(\frac{3}{2}\sin\omega_s t\cos\vartheta_k - \frac{3}{2}\cos\omega_s t\sin\vartheta_k\right) \\ &= k_d U_{max}\frac{3}{2}(\sin\omega_s t\cos\vartheta_k - \cos\omega_s t\sin\vartheta_k) = k_d U_{max}\frac{3}{2}\sin(\omega_s t - \vartheta_k). \end{aligned} \tag{196}$$

In the transients, if the speed is changing, the angle ϑ_k is given by Eq. (191). In the steady-state condition, when the speed is constant, ω_k = const., the equation for the voltage is as follows:

$$u_d = k_d U_{max}\frac{3}{2}\sin(\omega_s t - \omega_k t - \vartheta_{k0}) = k_d U_{max}\frac{3}{2}\sin((\omega_s - \omega_k)t - \vartheta_{k0}). \tag{197}$$

Here it is seen that the voltage in d-axis is alternating sinusoidal variable with the frequency which is the difference of the both systems: original three-phase abc system with the angular frequency ω_s and k-system, rotating with the speed ω_k.

Now the same approach will be used for the q-axis:

$$u_q = -k_q U_{max}\left(\sin\omega_s t\sin\vartheta_k + \sin\left(\omega_s t - \frac{2\pi}{3}\right)\sin\left(\vartheta_k - \frac{2\pi}{3}\right) + \sin\left(\omega_s t + \frac{2\pi}{3}\right)\sin\left(\vartheta_k + \frac{2\pi}{3}\right)\right). \tag{198}$$

The adjusting will result in equation:

$$u_q = -k_q U_{max}\frac{3}{2}(\sin\omega_s t\sin\vartheta_k + \cos\omega_s t\cos\vartheta_k),$$

which is finally accommodated to the form:

$$u_q = -k_q U_{max}\frac{3}{2}\cos(\omega_s t - \vartheta_k). \tag{199}$$

The voltage in the q-axis is shifted from the voltage in the d-axis about 90°, which is in coincidence with the definition of the d-axis and q-axis positions, which are perpendicular to each other. In transients when the speed is quickly changing, the angle ϑ_k is given by Eq. (191). In the steady-state condition, when the speed is constant, ω_k = const., the equation for the voltage is in the form:

$$u_{\mathrm{q}} = -k_{\mathrm{q}} U_{\max} \frac{3}{2} \cos\left(\omega_{\mathrm{s}} t - \omega_{\mathrm{k}} t - \vartheta_{\mathrm{k}0}\right) = -k_{\mathrm{q}} U_{\max} \frac{3}{2} \cos\left((\omega_{\mathrm{s}} - \omega_{\mathrm{k}}) t - \vartheta_{\mathrm{k}0}\right). \tag{200}$$

Finally, the equation for the zero component is adjusted as follows:

$$u_0 = k_0 U_{\max} \left(\sin \omega_{\mathrm{s}} t + \sin\left(\omega_{\mathrm{s}} t - \frac{2\pi}{3}\right) + \sin\left(\omega_{\mathrm{s}} t + \frac{2\pi}{3}\right)\right). \tag{201}$$

It is the sum of the voltage instantaneous values of the symmetrical three-phase system, which is, as it is known immediately, zero, or it is necessary to multiply all expressions for goniometrical functions, and after summarization of the appropriate members, the result is zero:

$$u_0 = k_0 U_{\max} \left(\sin \omega_{\mathrm{s}} t + \sin\left(\omega_{\mathrm{s}} t - \frac{2\pi}{3}\right) + \sin\left(\omega_{\mathrm{s}} t + \frac{2\pi}{3}\right)\right) = 0, \tag{202}$$

which is in coincidence with a note that the sum of the instantaneous values of variables, therefore also voltages, of the symmetrical three-phase system, is zero.

If the investigated three-phase system is not symmetrical, the zero component would have no zero value and would be necessary to add the equation for zero component to the dq0 system of equations. After the solution of dq0 variables, it would be necessary to make an inverse transformation on the basis of Eqs. (133)–(135), where component x_0 would appear.

Here the universality of the method of transformation is seen, because it is possible to investigate also unsymmetrical three-phase systems.

At the end of this chapter, the properties of the transformed sinusoidal variables are summarized, as shown in the above equations:

1. Variables d and q are alternating variables with a frequency which is given by the difference of the frequency of both systems: original three-phase system abc with the angular frequency ω_{s} and k-system rotating by angular speed ω_{k}.

2. Transformed variables d and q are shifted about 90°, unlike the three-phase system, in which the axes are shifted about 120°.

3. Variables of the zero component, i.e., with the subscript 0, are in the case of the symmetrical system, zero. If the three-phase system is not symmetrical, it is necessary to take the zero component into account, to find its value and to employ it in the inverse transformation into the system abc.

4. Magnitudes of variables dq0 depend on the choice of the constant of the proportionality.

The voltages in d-axis and q-axis are adjusted to the form:

$$u_{\mathrm{d}} = k_{\mathrm{d}} U_{\max} \frac{3}{2} \sin\left(\omega_{\mathrm{s}} t - \vartheta_{\mathrm{k}}\right) = U_{\mathrm{dmax}} \sin\left(\omega_{\mathrm{s}} t - \vartheta_{\mathrm{k}}\right), \tag{203}$$

whereby

$$U_{\mathrm{dmax}} = k_{\mathrm{d}} U_{\max} \frac{3}{2}, \tag{204}$$

$$u_q = -k_q U_{max} \frac{3}{2} \cos(\omega_s t - \vartheta_k) = -U_{qmax} \cos(\omega_s t - \vartheta_k), \tag{205}$$

whereby

$$U_{qmax} = k_q U_{max} \frac{3}{2}. \tag{206}$$

Here it is seen that if:

a. $k_d = k_q = \frac{2}{3}$, then $U_{dmax} = U_{qmax} = U_{max}$, but the principle of the power invariance is not valid.

b. $k_d = k_q = \sqrt{\frac{2}{3}}$, then $U_{dmax} = U_{qmax} = \sqrt{\frac{3}{2}} U_{max}$, but the principle of the power invariance is valid (see Section 8 and Eqs. (186) and (187)).

Equations for the voltages u_d and u_q are adjusted to the final form not only on the basis of the constants of proportionality but also on the basis of the k-system position, i.e., how the angle ϑ_k is chosen (see Section 10).

Note that if there are supposed cosinusoidal functions of the three-phase system, i.e.,

$$u_a = U_{max} \cos \omega_s t, \tag{207}$$

$$u_b = U_{max} \cos\left(\omega_s t - \frac{2\pi}{3}\right), \tag{208}$$

$$u_c = U_{max} \cos\left(\omega_s t + \frac{2\pi}{3}\right), \tag{209}$$

after the same approach at derivation as for sinusoidal functions, equations for the variables in d-axis and q-axis are gotten:

$$u_d = k_d U_{max} \frac{3}{2} \cos(\omega_s t - \vartheta_k), \tag{210}$$

$$u_q = k_q U_{max} \frac{3}{2} \sin(\omega_s t - \vartheta_k). \tag{211}$$

As it will be shown in Section 19, this version of the voltage origin of the three-phase system definition is more suitable for a synchronous machine because of the investigation of the load angle.

2.10 Choice of the angle ϑ_k and of the reference k-system position

The final form of the voltage equations in the system dq0 does not depend only on the choice of the proportionality constants but also on the position of the reference k-system and the angle ϑ_k and the angular speed ω_k.

The k-system can be positioned totally arbitrary, but some of the choices bring some simplicity in the investigation, which can be employed with benefit. Here are some of the most used possibilities, which are marked with special subscripts.

1. $\vartheta_k = 0, \omega_k = 0$, subscripts $\alpha, \beta, 0$.

This choice means that the k-system is identified with the axis of the stator a-phase winding, i.e., the k-system is static and does not rotate, much like stator a-phase winding.

This choice is distinguished from all others by subscripts. Instead of the subscripts d, q, 0, the subscripts $\alpha, \beta, 0$ are employed. Initially the subscripts α, β, γ were introduced, but after some development the new system of the subscripts $\alpha, \beta, 0$ was introduced because it was more logical. The zero component is the same as in the system dq0.

Equations for the voltages and currents are adjusted to be able to see how this choice brings benefits. From the equation in the previous chapter:

$$u_{\mathrm{d}} = k_{\mathrm{d}} U_{\max} \frac{3}{2} \sin\left(\omega_{\mathrm{s}} t - \vartheta_{\mathrm{k}}\right) = U_{\mathrm{dmax}} \sin\left(\omega_{\mathrm{s}} t - \vartheta_{\mathrm{k}}\right). \tag{212}$$

It is seen that if, simultaneously with the choice $\vartheta_{\mathrm{k}} = 0$, we take the proportional constants $k_{\mathrm{d}} = k_{\mathrm{q}} = \frac{2}{3}$ and change the subscripts, then for the original voltage, u_{d} is gotten:

$$u_{\alpha} = \frac{2}{3} U_{\max} \frac{3}{2} \sin\left(\omega_{\mathrm{s}} t - 0\right) = U_{\max} \sin \omega_{\mathrm{s}} t = u_{\mathrm{a}}, \tag{213}$$

which are very important findings, in that an instantaneous value of the voltage (and current) in the transformed system is identical with the instantaneous value of the voltage (or current) in phase a. This brings very simple situation, because it is not needed to make any inverse transformation.

Have a look at the voltage in the β-axis. According to the equation from the previous chapter for the u_{q}, and some accommodations, it results in the form:

$$u_{\beta} = -\frac{2}{3} U_{\max} \frac{3}{2} \cos\left(\omega_{\mathrm{s}} t - 0\right) = -U_{\max} \cos \omega_{\mathrm{s}} t, \tag{214}$$

which means that this voltage is fictitious and such voltage does not exist in the real abc system and is shifted about 90° from the voltage u_{α}.

It is the most important thing that in the same way the currents are transformed. It means that in the motoring operation, where the currents, together with the speed, are unknown, $i_{\alpha} = i_{\mathrm{a}}$ is gotten, which means that the transformed system solution brings directly the current in a-phase and no inverse transformation is needed. The currents in the rest of two phases b and c are shifted about 120°, if there is a symmetrical system. In such system, it is valid that the zero component is zero. If there is an unsymmetrical system, where zero component is not zero, all variables must be investigated in great details and to find the real values in the abc system by inverse transformation.

Additionally here are equations for an electromagnetic torque and time varying of the angular speed. On the basis of Eqs. (179) and (180), derived for the $k_{\mathrm{d}} = k_{\mathrm{q}} = \frac{2}{3}$, after the changing of the subscripts, the following is gained:

$$t_{\mathrm{e}} = p \frac{3}{2} \left(\psi_{\alpha} i_{\beta} - \psi_{\beta} i_{\alpha}\right), \tag{215}$$

$$\frac{\mathrm{d}\omega}{\mathrm{d}t} = \frac{p}{J} \left(p \frac{3}{2} \left(\psi_{\alpha} i_{\beta} - \psi_{\beta} i_{\alpha}\right) - t_{\mathrm{L}}\right). \tag{216}$$

At the end it is necessary to say that this choice is not profitable only for the squirrel cage asynchronous motors (see Section 11) but also for asynchronous motors with wound rotor and for asynchronous generators.

2. $\vartheta_k = \vartheta_r, \omega_k = \omega_r$, subscripts d, q, 0

This choice means that the k-system is identified with the rotor axis and the speed of the k-system with the rotor speed.

This transformation is employed with benefit for synchronous machines, because in equations for the voltage, there is a so-called load angle (see Eq. (218)), which is a very important variable in the operation of the synchronous machines. On the rotor of the synchronous machine, there is a concentrating field winding fed by DC current, which creates DC magnetic flux. Here the d-axis is positioned. Therefore the rotor system is not necessary to transform because the field winding is positioned directly in d-axis, and if the rotor has damping windings, they are decomposed into two axes, d-axis and q-axis, perpendicular to each other. Finally, as it was mentioned before, this transformation was developed for synchronous machine with salient poles; therefore, the subscripts d, q are left in the form, in which they were used during the whole derivation.

As in previous case, equations for the voltages and currents are again adjusted to be able to see advantage of this choice.

$$u_d = k_d U_{max} \frac{3}{2} \sin(\omega_s t - \vartheta_k) = U_{dmax} \sin(\omega_s t - \vartheta_k), \tag{217}$$

is seen that if $\vartheta_k = \vartheta_r$ is chosen, the argument of the sinusoidal function $(\omega_s t - \vartheta_r)$ is in fact the difference between the axis of rotating magnetic field and rotor position. This value is in the theory of synchronous machines defined as the load angle ϑ_L:

$$(\omega_s t - \vartheta_r) = \vartheta_L. \tag{218}$$

In Section 16 and 18 there will be derived, why in the case of synchronous machines the proportionality constants are chosen in this form:

$k_d = k_q = \sqrt{\frac{2}{3}}, k_0 = \sqrt{\frac{1}{3}}$.

Then the original voltage u_d in this system is in the form:

$$u_d = \sqrt{\frac{2}{3}} U_{max} \frac{3}{2} \sin(\omega_s t - \vartheta_r) = \sqrt{\frac{3}{2}} U_{max} \sin(\omega_s t - \vartheta_r), \tag{219}$$

and the voltage u_q after some accommodation is:

$$u_q = -\sqrt{\frac{2}{3}} U_{max} \frac{3}{2} \cos(\omega_s t - \vartheta_r) = -\sqrt{\frac{3}{2}} U_{max} \cos(\omega_s t - \vartheta_r). \tag{220}$$

Equations for the electromagnetic torque and time varying of the speed (181) and (182), derived for $k_d = k_q = \sqrt{\frac{2}{3}}$, are directly those equations, which are valid here, because the subscripts are not changed:

$$t_e = p\left(\psi_d i_q - \psi_q i_d\right), \tag{221}$$

$$\frac{d\omega}{dt} = \frac{p}{J}\left(p\left(\psi_d i_q - \psi_q i_d\right) - t_L\right). \tag{222}$$

Section 16 and others will deal with the synchronous machines in the general theory of electrical machines. These equations will be applied at the investigation of the properties of the synchronous machines in concrete examples, and also synchronous machines with permanent magnets will be investigated.

$$3.\ \vartheta_k = \int_0^t \omega_k dt + \vartheta_{k0} \tag{223}$$

$\omega_k = \omega_s$, subscripts x,y,0

This choice means that the speed of k-system rotation ω_k is identified with the synchronous speed of rotating magnetic field ω_s, i.e., transformation axes rotate with the same speed as the space vector of the stator voltages.

Adjust equations for the voltages, in which the advantage of this choice will be visible. From the equations in the previous chapter:

$$u_d = k_d U_{max} \frac{3}{2} \sin(\omega_s t - \vartheta_k) = U_{dmax} \sin(\omega_s t - \vartheta_k), \tag{224}$$

$$u_d = k_d U_{max} \frac{3}{2} \sin(\omega_s t - \omega_k t - \vartheta_{k0}) = k_d U_{max} \frac{3}{2} \sin((\omega_s - \omega_k)t - \vartheta_{k0}), \tag{225}$$

result that if ω_k introduces ω_s and changes the subscripts, the expression for u_d is in the form:

$$u_x = k_d U_{max} \frac{3}{2} \sin((\omega_s - \omega_s)t - \vartheta_{k0}) = k_d U_{max} \frac{3}{2} \sin(-\vartheta_{k0}) = -k_d U_{max} \frac{3}{2} \sin \vartheta_{k0}. \tag{226}$$

Similarly, for voltage u_q at the changed subscripts, the following is gained:

$$\begin{aligned} u_y &= -k_q U_{max} \frac{3}{2} \cos((\omega_s - \omega_s)t - \vartheta_{k0}) = -k_q U_{max} \frac{3}{2} \cos(-\vartheta_{k0}) \\ &= -k_q U_{max} \frac{3}{2} \cos \vartheta_{k0}. \end{aligned} \tag{227}$$

It is seen that both voltages in this system are constant DC variables, and it depends on the choice of the constants and initial value ϑ_{k0} which value they will have. At the suitable initial position of the transformation axes, one of them can be zero.

If, for example, the initial position of the k-system is chosen to be zero, $\vartheta_{k0} = 0$, and constants of proportionality $k_d = k_q = \frac{2}{3}$; then equations are very simplified and are as follows:

$$u_x = -k_d U_{max} \frac{3}{2} \sin(\vartheta_{k0}) = -\frac{2}{3} U_{max} \frac{3}{2} \sin 0 = 0, \tag{228}$$

$$u_y = -k_q U_{max} \frac{3}{2} \cos(\vartheta_{k0}) = -\frac{2}{3} U_{max} \frac{3}{2} \cos 0 = -U_{max}. \tag{229}$$

If it looks uncomfortable that both voltages are negative values, it is enough, if derivation of transformation equations from abc to dq0 start with an assumption that:

$$u_a = -U_{max} \sin \omega_s t, \tag{230}$$

$$u_b = -U_{max} \sin\left(\omega_s t - \frac{2\pi}{3}\right), \tag{231}$$

$$u_c = -U_{max} \sin\left(\omega_s t + \frac{2\pi}{3}\right). \tag{232}$$

In the steady-state condition, all variables on the stator and rotor are illustrated as DC variables. Therefore, the solution is very easy, but it is true that it is necessary to make an inverse transformation into the real abc system. This transformation system is very suitable for asynchronous motors.

The equation for torque is also very simplified, because the x-component of the current is also zero ($i_x = 0$). Then together with the change of the subscripts, Eq. (174) for torque, where the constants of the proportionality $k_d = k_q = \frac{2}{3}$ were used, is as follows:

$$t_e = p\frac{3}{2}\left(\psi_x i_y - \psi_y i_x\right) = p\frac{3}{2}\psi_x i_y, \tag{233}$$

and equation for time varying of the speed is:

$$\frac{d\omega}{dt} = \frac{p}{J}\left(p\frac{3}{2}\psi_x i_y - t_L\right). \tag{234}$$

4. ϑ_k, ω_k, are chosen totally generally, the position of the k-system is chosen totally generally, subscripts u, v, 0.

Although the whole derivation of transformed variables was made for the dq0 axis, because it was historically developed in such a way, and then the new subscripts were introduced by means of the special choice of the reference k-system position, it is seen that the subscripts dq0 are kept only for the synchronous machine, for which this transformation was developed. If it should be started now, perhaps two perpendicular axes to each other would be marked as u, v, 0. Nevertheless the original configuration of universal machine had windings in the axes d, q, and it is kept also for the future. However here introduced marking was not accepted by all experts dealing with this topic, and some authors used the system x, y, 0 instead of α, β, 0.

2.11 Asynchronous machine and its inductances

It is supposed that a reader is familiar with the basic design of asynchronous machine and its theory and properties. Now we will analyze the three-phase symmetrical system on the stator, marked abc and on the rotor, marked ABC, i.e., six windings together (**Figure 26**).

Basic voltage equations for the terminal voltage can be written for each winding or by one equation, at which the subscripts will be gradually changed:

$$u_j = R_j i_j + \frac{d\psi_j}{dt}, \tag{235}$$

where j = abc, ABC.

If the system is symmetrical, then it is possible to suppose that:

$$R_a = R_b = R_c = R_s, \tag{236}$$

$$R_A = R_B = R_C = R_r. \tag{237}$$

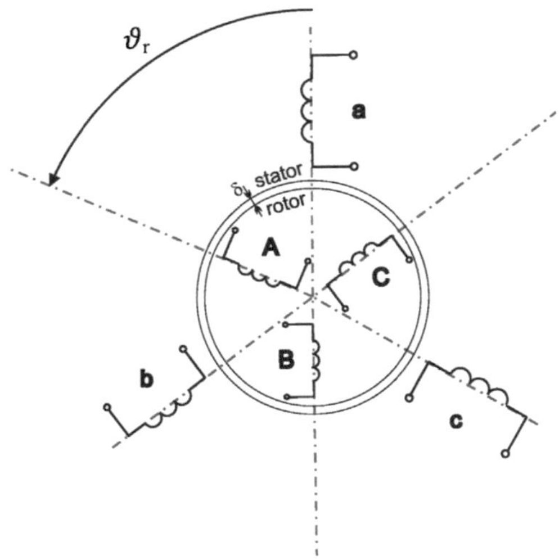

Figure 26.
Illustration figure of an asynchronous machine with three windings abc on the stator and three phase ABC on the rotor. They are shifted from each other about the angle ϑ_r.

Linkage magnetic flux can be also expressed by one equation as a sum of all winding contributions:

$$\psi_j = \sum_k \psi_{j,k} = \sum_k L_{j,k} i_k \tag{238}$$

where j, k = abc, ABC, but because of transparency here is the whole sum of the members in details:

$$\begin{aligned}
\psi_a &= L_{aa}i_a + L_{ab}i_b + L_{ac}i_c + L_{aA}i_A + L_{aB}i_B + L_{aC}i_C,\\
\psi_b &= L_{ba}i_a + L_{bb}i_b + L_{bc}i_c + L_{bA}i_A + L_{bB}i_B + L_{bC}i_C,\\
\psi_c &= L_{ca}i_a + L_{cb}i_b + L_{cc}i_c + L_{cA}i_A + L_{cB}i_B + L_{cC}i_C,\\
\psi_A &= L_{Aa}i_a + L_{Ab}i_b + L_{Ac}i_c + L_{AA}i_A + L_{AaB}i_B + L_{AC}i_C,\\
\psi_B &= L_{Ba}i_a + L_{Bb}i_b + L_{Bc}i_c + L_{BA}i_A + L_{BB}i_B + L_{BC}i_C,\\
\psi_C &= L_{Ca}i_a + L_{Cb}i_b + L_{Cc}i_c + L_{CA}i_A + L_{CB}i_B + L_{CC}i_C,
\end{aligned} \tag{239}$$

where:

$$L_{aa} = L_{bb} = L_{cc} = L_s \text{ are self} - \text{inductances of the stator windings.} \tag{240}$$

$$L_{AA} = L_{BB} = L_{CC} = L_r \text{ are self} - \text{inductances of the rotor windings.} \tag{241}$$

$$\begin{aligned} L_{ab} &= L_{ac} = L_{ba} = L_{bc} = L_{ca} = L_{cb}\\ &= -M_s \text{ are mutual inductances of the stator windings.}\end{aligned} \tag{242}$$

$$\begin{aligned} L_{AB} &= L_{AC} = L_{BA} = L_{BC} = L_{CA} = L_{CB}\\ &= -M_r \text{ are mutual inductances of the rotor windings.}\end{aligned} \tag{243}$$

The others are mutual inductances of stator and rotor windings. It is necessary to investigate if they depend on the rotor position or not.

2.11.1 Inductances that do not depend on the rotor position

1. Self-inductances of the stator windings L_s

DOI: http://dx.doi.org/10.5772/intechopen.91968

Self-inductance of stator single phase L_s without influence of the other stator phases and without influence of the rotor windings corresponds to the whole magnetic flux Φ_s, which is created by the single stator phase.

This flux is divided into two parts: leakage magnetic flux $\Phi_{\sigma s}$, which is linked only with the winding by which it was created and thus embraces only this phase, and magnetizing magnetic flux Φ_μ, which crosses air gap and enters the rotor and eventually is closed around the other stator or rotor windings. Inductances correspond with these fluxes according the permeance of the magnetic path and winding positions. Therefore it can be written:

$$\Phi_s = \Phi_{\sigma s} + \Phi_\mu \tag{244}$$

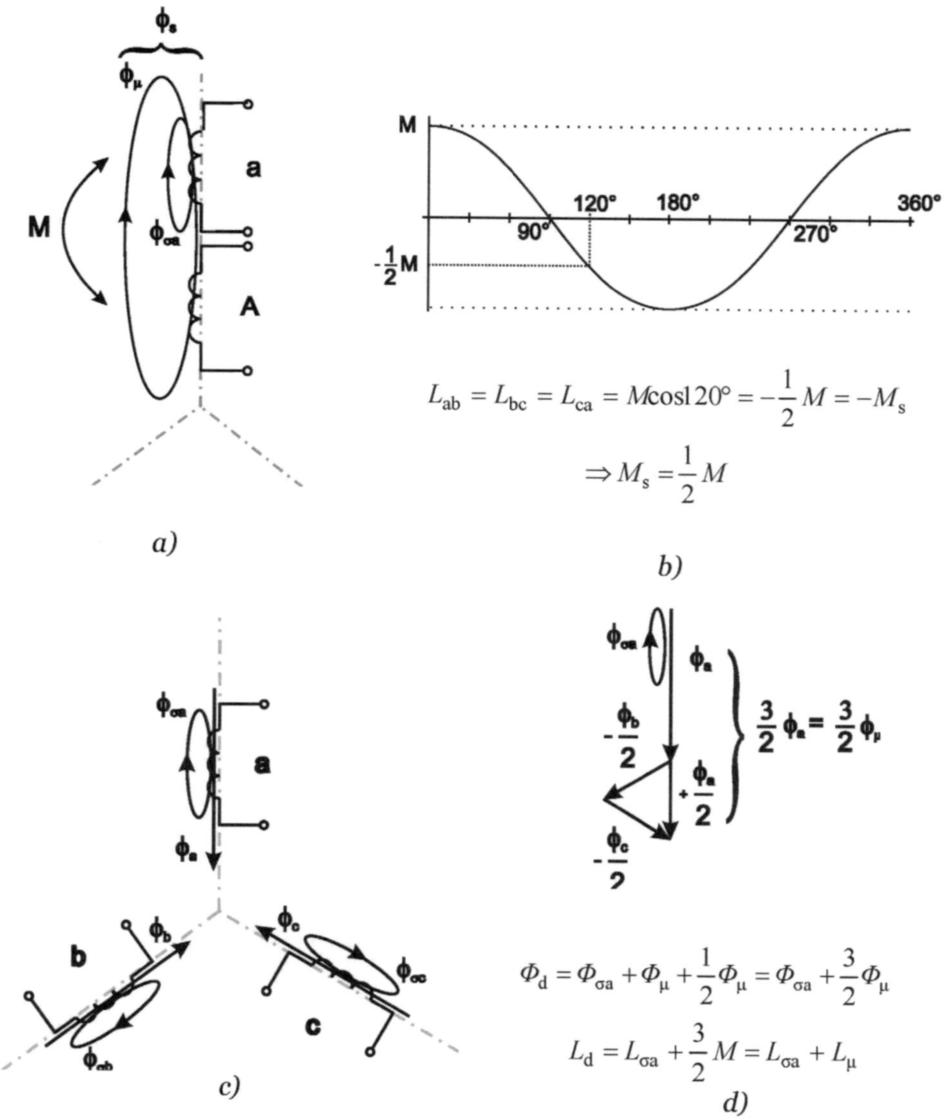

Figure 27.
(a) Illustration of mutual inductance M of the stator and rotor phase if their axes are identical, (b) justification of the mutual inductance value of the stator windings shifted about 120°, (c) magnetic flux directions if the phases are fed independently by positive currents, (d) sum of the magnetic fluxes if the windings are fed by three phases at the instant when a-phase has a positive magnitude and the b- and c-phases have half of negative magnitude.

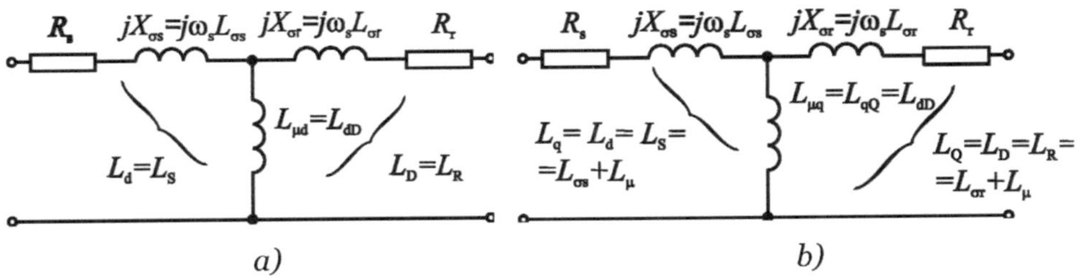

Figure 28.
Equivalent circuit of the asynchronous machine (a) in d-axis and (b) in q-axis. All rotor variables are referred to the stator.

$$L_s = L_{\sigma s} + M \tag{245}$$

where M is the mutual inductance of single stator phase and single rotor phase if their axes are identical (see **Figure 27a**).

2. Self-inductances of the rotor windings L_r.

These inductances are expressed similarly as the stator ones:

$$L_r = L_{\sigma r} + M. \tag{246}$$

Have a note that in the whole general theory of electrical machines, rotor variables are referred to the stator side.

3. Mutual inductances of the stator windings—M_s

Take an image that two stator windings have an identical axis, e.g., b-phase is identified with a-phase. Then their mutual inductance is M. Now the b-phase is moved to its original position, i.e., about 120°. According to **Figure 27b**, the value of the mutual inductance in this position is:

$$\cos 120° = -1/2\, M = -M_s \tag{247}$$

and this value is constant; it means it is always negative because the position of b-phase on the stator with regard to a-phase is stable.

4. Mutual inductances of the rotor windings—M_r.

The same analysis as in item 3 results in the finding that the mutual inductance of the rotor windings without the influence of the stator windings is always negative and equals (**Figure 28**).

$$-M_r = -\tfrac{1}{2} M \tag{248}$$

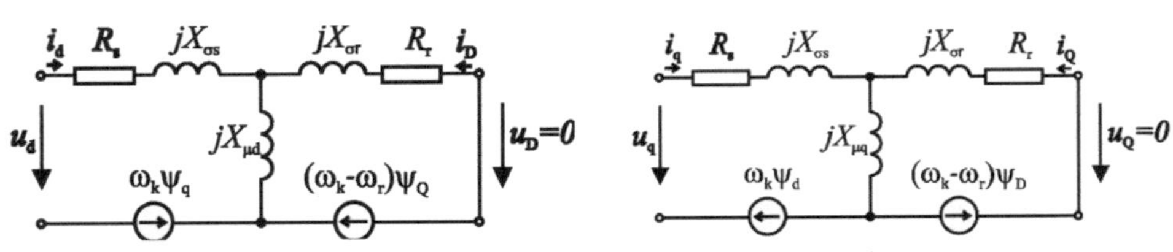

Figure 29.
Equivalent circuit of asynchronous machine for reference k-system: (a) in d-axis and (b) in q-axis for a chosen reference frame. All rotor variables are referred to the stator side to distinguish this case of transformation from the others, the axes are marked αβo (originally it was marked as αβγ), and also all subscripts of the currents, voltages, and linkage magnetic fluxes are with these subscripts. Then on the basis of the subscripts, it is possible to know what kind of transformation was used.

2.11.2 Inductances depending on the rotor position

All mutual inductances of the stator and rotor windings are expressed as follows:

$$L_{\mathrm{aA}} = L_{\mathrm{Aa}} = L_{\mathrm{bB}} = L_{\mathrm{Bb}} = L_{\mathrm{cC}} = L_{\mathrm{Cc}} = M\cos\vartheta_{\mathrm{r}},$$
$$L_{\mathrm{aB}} = L_{\mathrm{Ba}} = L_{\mathrm{bC}} = L_{\mathrm{Cb}} = L_{\mathrm{cA}} = L_{\mathrm{Ac}} = M\cos\left(\vartheta_{\mathrm{r}} + \frac{2\pi}{3}\right),$$
$$L_{\mathrm{aC}} = L_{\mathrm{Ca}} = L_{\mathrm{bA}} = L_{\mathrm{Ab}} = L_{\mathrm{cB}} = L_{\mathrm{Bc}} = M\cos\left(\vartheta_{\mathrm{r}} - \frac{2\pi}{3}\right), \tag{249}$$

where M is the mutual inductance of the stator and rotor phase if their axes are identical.

All expressions are introduced into (239); therefore, the inductances, linkage fluxes, and currents can be written in the matrix form:

$$L = \begin{bmatrix} L_s & -M_s & -M_s & M\cos\vartheta_r & M\cos\left(\vartheta_r + \frac{2}{3}\pi\right) & M\cos\left(\vartheta_r - \frac{2}{3}\pi\right) \\ -M_s & L_s & -M_s & M\cos\left(\vartheta_r - \frac{2}{3}\pi\right) & M\cos\vartheta_r & M\cos\left(\vartheta_r + \frac{2}{3}\pi\right) \\ -M_s & -M_s & L_s & M\cos\left(\vartheta_r + \frac{2}{3}\pi\right) & M\cos\left(\vartheta_r - \frac{2}{3}\pi\right) & M\cos\vartheta_r \\ M\cos\vartheta_r & M\cos\left(\vartheta_r - \frac{2}{3}\pi\right) & M\cos\left(\vartheta_r + \frac{2}{3}\pi\right) & L_r & -M_r & -M_r \\ M\cos\left(\vartheta_r + \frac{2}{3}\pi\right) & M\cos\vartheta_r & M\cos\left(\vartheta_r - \frac{2}{3}\pi\right) & -M_r & L_r & -M_r \\ M\cos\left(\vartheta_r - \frac{2}{3}\pi\right) & M\cos\left(\vartheta_r + \frac{2}{3}\pi\right) & M\cos\vartheta_r & -M_r & -M_r & L_r \end{bmatrix}$$

$$\begin{bmatrix} \psi_a \\ \psi_b \\ \psi_c \\ \psi_A \\ \psi_B \\ \psi_C \end{bmatrix} = [L] \begin{bmatrix} i_a \\ i_b \\ i_c \\ i_A \\ i_B \\ i_C \end{bmatrix} \tag{250}$$

After these expressions are introduced into (235), six terminal voltage equations are obtained, with nonlinear, periodically repeated coefficients $M\cos\vartheta_{\mathrm{r}}$, $M\cos\left(\vartheta_{\mathrm{r}} - \frac{2\pi}{3}\right)$, and $M\cos\left(\vartheta_{\mathrm{r}} + \frac{2\pi}{3}\right)$, where M is the mutual inductance of the stator and rotor phase if their axes are identical and ϑ_{r} is an angle between the axis of the same stator and rotor phase (**Figure 29**).

To solve such equations is very complicated; therefore, it is necessary to eliminate the periodically repeated coefficients. This is possible to do by various real or complex linear transformations. The most employed is Park linear transformation, mentioned in Section 4. In the next it is applied for this case.

2.12 Linkage magnetic flux equations of the asynchronous machine in the general theory of electrical machines

On the basis of the equations for transformation into dq0 system, the equation for ψ_{d} is written:

$$\psi_{\mathrm{d}} = k_{\mathrm{d}}\left(\psi_{\mathrm{a}}\cos\vartheta_{\mathrm{k}} + \psi_{\mathrm{b}}\cos\left(\vartheta_{\mathrm{k}} - \frac{2\pi}{3}\right) + \psi_{\mathrm{c}}\cos\left(\vartheta_{\mathrm{k}} + \frac{2\pi}{3}\right)\right). \tag{251}$$

The expressions for ψ_a, ψ_b a ψ_c from (250) are introduced into it:

$$\begin{aligned}\psi_d = {} & k_d\left((L_s i_a - M_s i_b - M_s i_c)\cos\vartheta_k + \left(M(\cos\vartheta_r)i_A + M\cos\left(\vartheta_r + \frac{2\pi}{3}\right)i_B + M\cos\left(\vartheta_r - \frac{2\pi}{3}\right)i_C\right)(\cos\vartheta_k)\right)\\ & + k_d\left((-M_s i_a + L_s i_b - M_s i_c)\cos\left(\vartheta_k - \frac{2\pi}{3}\right) + \left(M\cos\left(\vartheta_r - \frac{2\pi}{3}\right)i_A + M\cos(\vartheta_r)i_B + M\cos\left(\vartheta_r + \frac{2\pi}{3}\right)i_C\right)\cos\left(\vartheta_k - \frac{2\pi}{3}\right)\right)\\ & + k_d\left((-M_s i_a - M_s i_b + L_s i_c)\cos\left(\vartheta_k + \frac{2\pi}{3}\right) + \left(M\cos\left(\vartheta_r + \frac{2\pi}{3}\right)i_A + M\cos\left(\vartheta_r - \frac{2\pi}{3}\right)i_B + M\cos(\vartheta_r)i_C\right)\cos\left(\vartheta_k + \frac{2\pi}{3}\right)\right)\end{aligned} \tag{252}$$

If we consider that:

$$\begin{aligned} i_0 &= k_0(i_a + i_b + i_c),\\ \frac{i_0}{k_0} &= (i_a + i_b + i_c),\\ i_b + i_c &= \frac{i_0}{k_0} - i_a,\\ -M_s(i_b + i_c) &= -M_s\left(\frac{i_0}{k_0} - i_a\right),\\ -M_s(i_a + i_c) &= -M_s\left(\frac{i_0}{k_0} - i_b\right),\\ -M_s(i_a + i_b) &= -M_s\left(\frac{i_0}{k_0} - i_c\right),\end{aligned}$$

and these expressions are introduced into the equation above for all three phases, after modifications, some expressions that are zero are found, e.g.:

$$-M_s\left(\frac{i_0}{k_0}\right)\left(\cos\vartheta_k + \cos\left(\vartheta_k - \frac{2\pi}{3}\right) + \cos\left(\vartheta_k + \frac{2\pi}{3}\right)\right) = 0, \tag{253}$$

and others in which transformed variables are seen, e.g.:

$$M_s k_d\left(i_a\cos\vartheta_k + i_b\cos\left(\vartheta_k - \frac{2\pi}{3}\right) + i_c\cos\left(\vartheta_k + \frac{2\pi}{3}\right)\right) = M_s i_d, \tag{254}$$

or:

$$L_s k_d\left(i_a\cos\vartheta_k + i_b\cos\left(\vartheta_k - \frac{2\pi}{3}\right) + i_c\cos\left(\vartheta_k + \frac{2\pi}{3}\right)\right) = L_s i_d. \tag{255}$$

If these two expressions are summed, it results in:

$$(L_s + M_s)i_d = L_d i_d, \tag{256}$$

where L_d is introduced as the sum of the self L_s and mutual inductance M_s of the stator windings in the d-axis. Then it is seen that L_d is a total inductance of the stator windings in the d-axis:

$$L_d = L_s + M_s = L_{\sigma s} + M + \frac{M}{2} = L_{\sigma s} + \frac{3}{2}M = L_{\sigma s} + L_{\mu d} = L_{\sigma s} + L_{\mu}. \tag{257}$$

This is evident also in **Figure 27a**, if a constant air gap of asynchronous machines is taken into account. Therefore, inductances in d-axis and q-axis are equal, and there is no need to mark separately magnetizing inductance in d-axis and q-axis. The members of Eq. (252), in which act rotor currents i_A, i_B, i_C, can be also accommodated in a similar way as the stator currents, which results in the following:

$$\frac{3}{2}Mk_d\left(i_A\cos(\vartheta_k-\vartheta_r)+i_B\cos\left(\vartheta_k-\vartheta_r-\frac{2\pi}{3}\right)+i_C\cos\left(\vartheta_k-\vartheta_r+\frac{2\pi}{3}\right)\right)=L_{dD}i_D. \tag{258}$$

In this expression there is used a knowledge, that: (1) mutual inductance of the stator and rotor winding with contribution of all three stator phases is 3/2 M what is marked L_{dD}, but it is known that in the equivalent circuit is marked as L_μ and (2) the angle between the axis of the rotor phase and the axis of the reference k-system is $(\vartheta_k-\vartheta_r)$. Therefore, the rotor variables are transformed into k-system by means of this angle.

Then it is possible to write that the transformed current of the rotor system is i_D:

$$k_d\left(i_A\cos(\vartheta_k-\vartheta_r)+i_B\cos\left(\vartheta_k-\vartheta_r-\frac{2\pi}{3}\right)+i_C\cos\left(\vartheta_k-\vartheta_r+\frac{2\pi}{3}\right)\right)=i_D, \tag{259}$$

and the whole Eq. (252) can be written much more briefly:

$$\psi_d=L_di_d+L_{dD}i_D, \tag{260}$$

where L_d is given by the (257) and

$$L_{dD}=\frac{3}{2}M=L_\mu. \tag{261}$$

The equation for ψ_q is obtained in a similar way and after accommodations is written in the form:

$$\psi_q=L_qi_q-\frac{3}{2}k_qM\left(i_A\sin(\vartheta_k-\vartheta_r)+i_B\sin\left(\vartheta_k-\vartheta_r-\frac{2\pi}{3}\right)+i_C\sin\left(\vartheta_k-\vartheta_r+\frac{2\pi}{3}\right)\right), \tag{262}$$

or briefly:

$$\psi_q=L_qi_q+L_{qQ}i_Q, \tag{263}$$

where:

$$L_{qQ}=\frac{3}{2}M=L_\mu, \tag{264}$$

and

$$-k_q\left(i_A\sin(\vartheta_k-\vartheta_r)+i_B\sin\left(\vartheta_k-\vartheta_r-\frac{2\pi}{3}\right)+i_C\sin\left(\vartheta_k-\vartheta_r+\frac{2\pi}{3}\right)\right)=i_Q. \tag{265}$$

Considering that in the asynchronous machine the air gap is constant around the whole periphery of the stator boring, there is no difference in the inductances in d-axis and q-axis; therefore, the following can be written:

$$\psi_q = L_d i_q + L_{dD} i_Q, \tag{266}$$

and also

$$L_{dD} = L_{qQ} = \frac{3}{2} M = L_\mu. \tag{267}$$

Zero component is as follow:

$$\psi_0 = L_0 i_0, \tag{268}$$

where:

$$L_0 = L_s - 2M_s = L_{\sigma s} + M - 2\frac{M}{2} = L_{\sigma s}. \tag{269}$$

The fact that the zero-component inductance L_0 is equal to the stator leakage inductance $L_{\sigma s}$ can be used with a benefit if $L_{\sigma s}$ should be measured. All three phases of the stator windings are connected together in a series, or parallelly, and fed by a single-phase voltage. In this way a pulse, non-rotating, magnetic flux is created. Thus a zero, non-rotating, component of the voltage, current, and impedance is measured.

Linear transformation is employed also at rotor linkage magnetic flux derivations in the system DQ0:

$$\psi_D = L_D i_D + L_{Dd} i_d, \tag{270}$$

$$\psi_Q = L_Q i_Q + L_{Qq} i_q, \tag{271}$$

eventually considering that the air gap is constant and the parameters in the d-axis and q-axis are equal:

$$\psi_Q = L_D i_Q + L_{Dd} i_q. \tag{272}$$

The meaning of the rotor parameters is as follows:

$$L_D = L_Q = L_r + M_r = L_{\sigma r} + M + \frac{M}{2} = L_{\sigma r} + \frac{3}{2} M = L_{\sigma r} + L_\mu. \tag{273}$$

Similarly, for the zero rotor component can be written as:

$$\psi_O = L_O i_O, \tag{274}$$

where:

$$L_O = L_r - 2M_r = L_{\sigma r} + M - 2\frac{M}{2} = L_{\sigma r}. \tag{275}$$

Take into account that all rotor variables are referred to the stator side; eventually they are measured from the stator side.

2.13 Voltage equations of the asynchronous machine after transformation into k-system with d-axis and q-axis

Voltage equations of the asynchronous machines in the dq0 system are obtained by a procedure described in Section 7. There are equations for the stator terminal voltage in the form:

$$u_{\mathrm{d}} = R_{\mathrm{s}} i_{\mathrm{d}} + \frac{\mathrm{d}\psi_{\mathrm{d}}}{\mathrm{d}t} - \omega_{\mathrm{k}} \psi_{\mathrm{q}}, \tag{276}$$

$$u_{\mathrm{q}} = R_{\mathrm{s}} i_{\mathrm{q}} + \frac{\mathrm{d}\psi_{\mathrm{q}}}{\mathrm{d}t} + \omega_{\mathrm{k}} \psi_{\mathrm{d}}, \tag{277}$$

$$u_{0} = R_{\mathrm{s}} i_{0} + \frac{\mathrm{d}\psi_{0}}{\mathrm{d}t}. \tag{278}$$

The rotor voltage equations are derived in a similar way as the stator ones but with a note that the rotor axis is shifted from the k-system axis about the angle $(\vartheta_{\mathrm{k}} - \vartheta_{\mathrm{r}})$; thus in the equations there are members with the angular speed $(\omega_{\mathrm{k}} - \omega_{\mathrm{r}})$:

$$u_{\mathrm{D}} = R_{\mathrm{r}} i_{\mathrm{D}} + \frac{\mathrm{d}\psi_{\mathrm{D}}}{\mathrm{d}t} - (\omega_{\mathrm{k}} - \omega_{\mathrm{r}}) \psi_{\mathrm{Q}}, \tag{279}$$

$$u_{\mathrm{Q}} = R_{\mathrm{r}} i_{\mathrm{Q}} + \frac{\mathrm{d}\psi_{\mathrm{Q}}}{\mathrm{d}t} + (\omega_{\mathrm{k}} - \omega_{\mathrm{r}}) \psi_{\mathrm{D}}, \tag{280}$$

$$u_{\mathrm{O}} = R_{\mathrm{r}} i_{\mathrm{O}} + \frac{\mathrm{d}\psi_{\mathrm{O}}}{\mathrm{d}t}. \tag{281}$$

These six equations create a full system of the asynchronous machine voltage equations. Rotor variables are referred to the stator side; expressions for the linkage magnetic flux are shown in Section 12.

2.14 Asynchronous motor and its equations in the system αβ0

According to Sections 7 and 10, the reference k-system can be positioned arbitrarily, but some specific positions can simplify solutions; therefore, they are used with a benefit. One of such cases happens if the d-axis of the k-system is identified with the axis of the stator a-phase; it means $\vartheta_k = 0, \omega_k = 0$. This system is in this book marked as αβ0 system.

This system is obtained by phase variable projection into stationary reference system, linked firmly with a-phase. It is a two-axis system, and zero components are identical with the non-rotating components known from the theory of symmetrical components.

The original voltage equations of asynchronous machine derived in Sections 7 and 13 are as follows:

$$u_{\mathrm{d}} = R_{\mathrm{s}} i_{\mathrm{d}} + \frac{\mathrm{d}\psi_{\mathrm{d}}}{\mathrm{d}t} - \omega_{\mathrm{k}} \psi_{\mathrm{q}}, \tag{282}$$

$$u_{\mathrm{q}} = R_{\mathrm{s}} i_{\mathrm{q}} + \frac{\mathrm{d}\psi_{\mathrm{q}}}{\mathrm{d}t} + \omega_{\mathrm{k}} \psi_{\mathrm{d}}, \tag{283}$$

$$u_{0} = R_{\mathrm{s}} i_{0} + \frac{\mathrm{d}\psi_{0}}{\mathrm{d}t}, \tag{284}$$

$$u_{\mathrm{D}} = R_{\mathrm{r}} i_{\mathrm{D}} + \frac{\mathrm{d}\psi_{\mathrm{D}}}{\mathrm{d}t} - (\omega_{\mathrm{k}} - \omega_{\mathrm{r}}) \psi_{\mathrm{Q}}, \tag{285}$$

$$u_{\mathrm{Q}} = R_{\mathrm{r}} i_{\mathrm{Q}} + \frac{\mathrm{d}\psi_{\mathrm{Q}}}{\mathrm{d}t} + (\omega_{\mathrm{k}} - \omega_{\mathrm{r}}) \psi_{\mathrm{D}}, \tag{286}$$

$$u_{\mathrm{O}} = R_{\mathrm{r}} i_{\mathrm{O}} + \frac{\mathrm{d}\psi_{\mathrm{O}}}{\mathrm{d}t}, \tag{287}$$

where:

$$\psi_d = L_d i_d + L_{dD} i_D, \tag{288}$$

$$\psi_q = L_q i_q + L_{qQ} i_Q, \tag{289}$$

$$\psi_0 = L_0 i_0, \tag{290}$$

$$L_d = L_q = L_{\sigma s} + L_\mu, \tag{291}$$

$$L_{dD} = L_{Dd} = \frac{3}{2}M = L_\mu, \tag{292}$$

$$L_{qQ} = L_{Qq} = L_{dD} = \frac{3}{2}M = L_\mu, \tag{293}$$

$$\psi_D = L_D i_D + L_{Dd} i_d, \tag{294}$$

$$\psi_Q = L_Q i_Q + L_{Qq} i_q, \tag{295}$$

$$\psi_O = L_O i_O, \tag{296}$$

$$L_D = L_Q = L_r + M_r = L_{\sigma r} + M + \frac{M}{2} = L_{\sigma r} + \frac{3}{2}M = L_{\sigma r} + L_\mu, \tag{297}$$

$$L_O = L_r - 2M_r = L_{\sigma r} + M - 2\frac{M}{2} = L_{\sigma r}. \tag{298}$$

Now new subscripts the following are introduced:
For currents and voltages:

$$\mathrm{d} = \alpha\mathrm{s}, \mathrm{q} = \beta\mathrm{s}, \mathrm{D} = \alpha\mathrm{r}, \mathrm{Q} = \beta\mathrm{r}, \tag{299}$$

For inductances:

$$L_d = L_q = L_{\sigma s} + L_\mu = L_S, \tag{300}$$

$$L_D = L_Q = L_{\sigma r} + L_\mu = L_R, \tag{301}$$

$$L_{dD} = L_{qQ} = \frac{3}{2}M = L_\mu. \tag{302}$$

The original equations, rewritten with the new subscripts, with the fact that $\vartheta_k = 0$ and $\omega_k = 0$ and with an assumption that the three-phase system is symmetrical, meaning the zero components are zero, are as follows:

$$u_{\alpha s} = R_s i_{\alpha s} + L_S \frac{di_{\alpha s}}{dt} + L_\mu \frac{di_{\alpha r}}{dt} \tag{303}$$

$$u_{\beta s} = R_s i_{\beta s} + L_S \frac{di_{\beta s}}{dt} + L_\mu \frac{di_{\beta r}}{dt} \tag{304}$$

$$u_{\alpha r} = R_r i_{\alpha r} + \omega_r L_R i_{\beta r} + \omega_r L_\mu i_{\beta s} + L_R \frac{di_{\alpha r}}{dt} + L_\mu \frac{di_{\alpha s}}{dt} \tag{305}$$

$$u_{\beta r} = R_r i_{\beta r} - \omega_r L_R i_{\alpha r} - \omega_r L_\mu i_{\alpha s} + L_R \frac{di_{\beta r}}{dt} + L_\mu \frac{di_{\beta s}}{dt} \tag{306}$$

If transients are solved for motoring operation, then stator terminal voltages on the left side of the equations are known variables and are necessary to introduce derived expressions for sinusoidal variables transformed into dq0 system, now α, β-axes ((196) for u_d and (199) for u_q). Rotor voltages are zero, if there is squirrel cage rotor. If there is wound rotor, here is a possibility to introduce a voltage applied to the rotor terminals, as in the case of asynchronous generator for wind power stations, where the armature winding is connected to the frequency converter. If the rotor winding is short circuited, then the rotor voltages are also zero.

In the motoring operation, the terminal voltages are known variables, and unknown variables are currents and speed. Therefore it is suitable to accommodate the previous equations in the form where the unknown variables are solved. From Eq. (303), the following is obtained:

$$L_S \frac{di_{\alpha s}}{dt} = u_{\alpha s} - R_s i_{\alpha s} - L_\mu \frac{di_{\alpha r}}{dt}, \tag{307}$$

and from Eq. (305):

$$\frac{di_{\alpha r}}{dt} = \frac{1}{L_R}\left(u_{\alpha r} - R_r i_{\alpha r} - \omega_r L_R i_{\beta r} - \omega_r L_\mu i_{\beta s} - L_\mu \frac{di_{\alpha s}}{dt}\right). \tag{308}$$

This equation is introduced into Eq. (307). Then it is possible to eliminate a time variation of the stator current in the α-axis:

$$\frac{di_{\alpha s}}{dt} = \frac{L_R}{L_S L_R - L_\mu^2}\left(u_{\alpha s} - R_s i_{\alpha s} + \frac{L_\mu}{L_R} R_r i_{\alpha r} + \omega_r \frac{L_\mu^2}{L_R} i_{\beta s} + \omega_r L_\mu i_{\beta r} - \frac{L_\mu}{L_R} u_{\alpha r}\right). \tag{309}$$

The same way is applied for the other current components:

$$\frac{di_{\alpha r}}{dt} = \frac{L_S}{L_S L_R - L_\mu^2}\left(u_{\alpha r} - R_r i_{\alpha r} + \frac{L_\mu}{L_S} R_s i_{\alpha s} - \omega_r L_\mu i_{\beta s} - \omega_r L_R i_{\beta r} - \frac{L_\mu}{L_S} u_{\alpha s}\right), \tag{310}$$

$$\frac{di_{\beta s}}{dt} = \frac{L_R}{L_S L_R - L_\mu^2}\left(u_{\beta s} - R_s i_{\beta s} + \frac{L_\mu}{L_R} R_r i_{\beta r} - \omega_r \frac{L_\mu^2}{L_R} i_{\alpha s} - \omega_r L_\mu i_{\alpha r} - \frac{L_\mu}{L_R} u_{\beta r}\right), \tag{311}$$

$$\frac{di_{\beta r}}{dt} = \frac{L_S}{L_S L_R - L_\mu^2}\left(u_{\beta r} - R_r i_{\beta r} + \frac{L_\mu}{L_S} R_s i_{\beta s} + \omega_r L_\mu i_{\alpha s} + \omega_r L_R i_{\alpha r} - \frac{L_\mu}{L_S} u_{\beta s}\right). \tag{312}$$

The last equation is for time variation of the speed. On the basis of Section 8, if in the equation for the electromagnetic torque the constants $k_d = k_q = 2/3$ are introduced and after changing the subscripts, the torque is in the form:

$$t_e = p\frac{2}{3}\frac{1}{k_d k_q}\left(\psi_d i_q - \psi_q i_d\right) = p\frac{3}{2}\left(\psi_{\alpha s} i_{\beta s} - \psi_{\beta s} i_{\alpha s}\right) = p\frac{3}{2}L_\mu\left(i_{\alpha r} i_{\beta s} - i_{\beta r} i_{\alpha s}\right),$$

$$t_e = p\frac{3}{2}L_\mu\left(i_{\alpha r} i_{\beta s} - i_{\beta r} i_{\alpha s}\right). \tag{313}$$

After considering Eq. (176), the electrical angular speed is obtained in the form:

$$\frac{d\omega_r}{dt} = \frac{p}{J}\left[p\frac{3}{2}L_\mu\left(i_{\alpha r} i_{\beta s} - i_{\beta r} i_{\alpha s}\right) - t_L\right]. \tag{314}$$

Mechanical angular speed is linked through the number of the pole pairs $\Omega_r = \frac{\omega_r}{p}$, which directly corresponds to the revolutions per minute.

For Eqs. (309)–(312), the next expressions are introduced for the voltages (see Sections 9 and 10):

$$u_{\alpha s} = U_m \sin \omega_s t = u_a, \tag{315}$$

$$u_{\beta s} = -U_m \cos \omega_s t, \tag{316}$$

which is displaced about 90° with regard to the $u_{\alpha s}$. Rotor voltages in the most simple case for the squirrel cage rotor are zero:

$$u_{\alpha r} = u_{\beta r} = 0. \tag{317}$$

In the next chapter, solving of the transients in a concrete asynchronous motor with squirrel cage rotor and wound rotor is shown.

2.15 Simulation of the transients in asynchronous motors

2.15.1 Asynchronous motor with squirrel cage rotor

Equations derived in the previous chapter are applied on a concrete asynchronous motor with squirrel cage rotor. The rotor bars are short circuited by end rings; thus the rotor voltages $u_{\alpha r}$ and $u_{\beta r}$ in Eqs. (309)–(312) are zero.

In **Figure 30**, simulation waveforms of the starting up of an asynchronous motor when it is switched directly across the line are shown. Parameters of the investigated motor are in **Table 3**.

Simulation waveforms in **Figure 30a–c** show time variations of the variables n = f(t), i_a = f(t), and t_e = f(t) after switching the motor directly across the line. At the instant t = 0.5 s, the motor is loaded by the rated torque T_N = 3.7 Nm. In

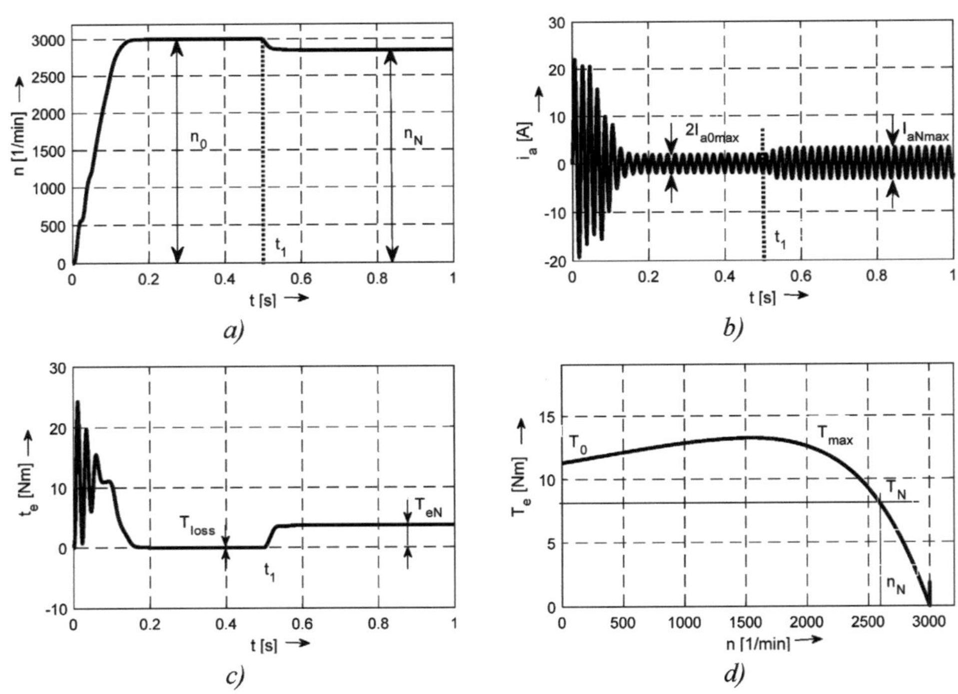

Figure 30.
Simulation waveforms of the asynchronous motor at its switching directly across the line, time waveforms of the (a) speed, (b) a-phase current I_a*, (c) developed electromagnetic torque, and (d) torque vs. speed if the motor is fed by rated voltage.*

P_N = 1.1 kW	R_s = 6.46 Ω
U_N = 230 V	$L_s = L_r$ = 0.5419 H
I_N = 2.4 A	R_r = 5.8 Ω
f_N = 50 Hz	L_μ = 0.5260 H
n_N = 2845 min^{-1}	J = 0.04 kg m^2
T_N = 3.7 Nm	p = 1
T_{loss} = 0.1 Nm	

Table 3.
Nameplate and parameters of the investigated asynchronous motor.

Figure 30d, torque vs. speed curve T_e = f(n) is shown. As it is seen from the waveforms, during the starting up, the motor develops very high starting torque, which could be dangerous for mechanical load of some parts of the drive system, and there are very high starting currents, which could be dangerous for the motor because of its heating and for the feeding part of the drive.

Relatively large starting current can cause an appreciable drop in motor terminal voltage, which reduces the starting current but also the starting torque. If the supply voltage drop would be excessive, some kind of across-the-line starter that reduces the terminal voltage and hence the starting current is required. For this purpose, a three-phase step-down autotransformer may be employed. The autotransformer is switched out of the circuit as the motor approaches full speed. The other method of starting is by a star-delta switch or by inserting resistances into the stator winding circuit. In the industry, a special apparatus is used, the so-called softstarter, which enables the starting of the defined requirement. Softstarter contains solid-state elements (thyristors), which enable to vary the terminal voltage of the motor. The start up is carried out by limitation of the maximal value of current, which will not be gotten over during the starting. This control is ensured by the possibility to change the terminal voltage of the motor. The more sophisticated way is a frequency starting during which not only voltage but also frequency is gradually increased, whereby the ratio U/f is kept constant. During start up, also maximum of the speed acceleration is defined.

Simulations of softstarter and frequency converter applications are shown in **Figure 31**. In both cases not only value of the starting torque is reduced, which is undesirable, but also the value of the starting current. The current does not cross the rated value and in this simulated case neither no-load current I_{a0}. It is seen in comparison waveforms in **Figure 30b** with waveforms in **Figure 31(c)** and **(d)**.

2.15.2 Asynchronous motor with wound rotor

Equations in Section 14 are the basis for the simulations. In this case, it is possible to feed the terminals of the wound armature on the rotor. This possibility is employed in applications with asynchronous generators, where feeding to the rotor serves as stabilization of the output frequency of the generator. Previously, the rotor terminals of the asynchronous motor were used for variation of the rotor circuit resistance by external rheostats. Such starting up is shown in this part. The nameplate and parameters of the investigated motor are in **Table 4**.

Simulations are shown in **Figure 32a–c**. There are time waveforms of the variables n = f(t), i_a = f(t), and t_e = f(t) after the switching directly across the line.

At the instant t = 0.5 s, the motor is loaded by the rated torque T_N = 30 Nm. In **Figure 32d** there is a curve T_e = f(n).

Simulation waveforms are very similar with those of the squirrel cage rotor (high starting current and torques). But in the case of wound rotor, there is a possibility to add external resistors and to control the current and the torque (**Figure 33**).

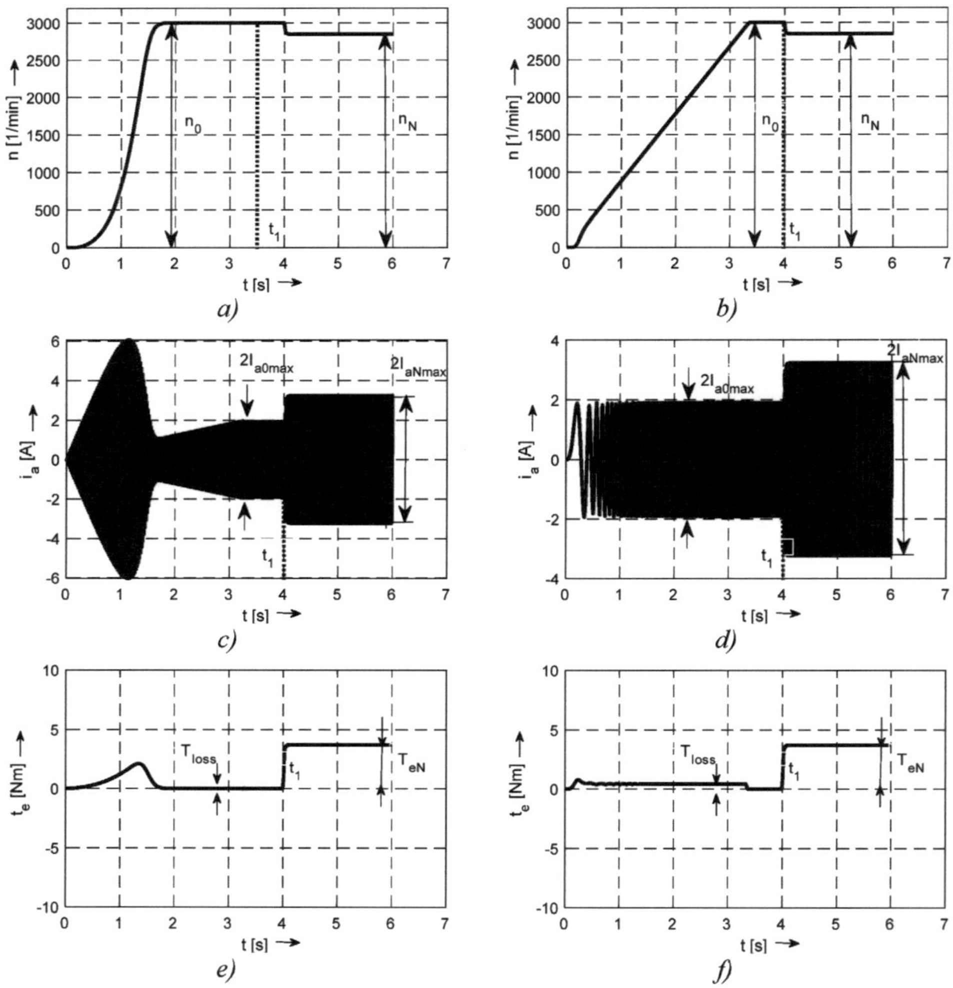

Figure 31.
Simulations of the asynchronous motor starting up by means of softstarter, time waveforms of (a) speed n*, (c) phase current* I_a*, and (e) developed electromagnetic torque, and by means of a frequency converter, again in the same order: time waveforms of the (b) speed* n*, (d) phase current* I_a*, and (f) developed electromagnetic torque.*

P_N = 4.4 kW	R_s = 1.125 Ω
U_{sN} = 230 V , U_{rN} = 64 V	L_s = L_r = 0.1419 H
I_{sN} = 9.4 A, I_{rN} = 47 A	R_r = 1.884 Ω
f_N = 50 Hz	L_μ = 0.131 H
n_N = 1370 min–^{1}	J = 0.04 kg m^2
T_N = 30 Nm	p = 2
T_{loss} = 0.1 Nm	

Table 4.
Nameplate and parameters of the investigated wound rotor asynchronous motor.

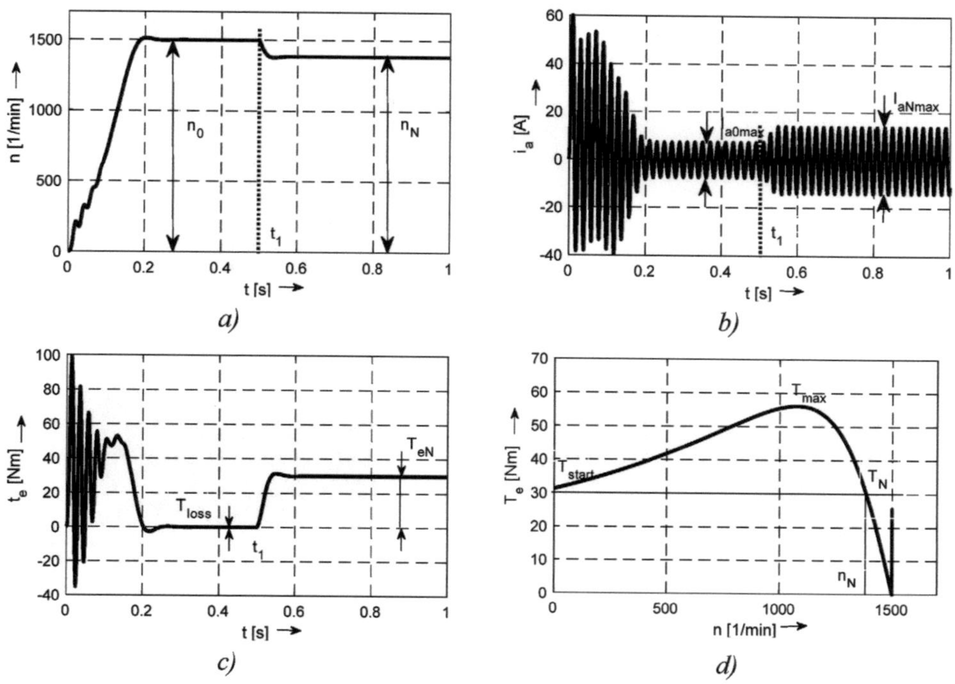

Figure 32.
Simulations of the rotor wound asynchronous motor during the switching directly across the line. Time waveforms of the (a) speed, (b) phase current I_a, (c) developed electromagnetic torque, and (d) torque vs. speed at rated voltage.

2.16 Synchronous machine and its inductances

It is supposed that a reader is familiar with the basic knowledge of a synchronous machine theory, properties, and design configuration. The synchronous machine with salient poles on the rotor; symmetrical three-phase system a, b, c on the stator; field winding f in the d-axis on the rotor; and damping winding, split into two parts perpendicular to each other (D and Q on the rotor), positioned in the d-axis and q-axis, as it is seen in **Figure 34**, is analyzed. The d-axis on the rotor is shifted about the angle ϑ_r from the axis of the a-phase on the stator.

Basic equations for terminal voltage can be written for each of the winding separately, or briefly by one equation, in which the subscripts are gradually changed for each winding:

$$u_j = R_j i_j + \frac{d\psi_j}{dt} \tag{318}$$

where j = a, b, c, f, D, Q.

If symmetrical three-phase winding on the stator is supposed, then it can be supposed that their resistances are identical and can be marked by the subscript "s":

$$R_a = R_b = R_c = R_s \tag{319}$$

Linkage magnetic flux can be also expressed briefly by the sum of all winding contributions:

$$\psi_j = \sum_k \psi_{j,k} = \sum_k L_{j,k} i_k \tag{320}$$

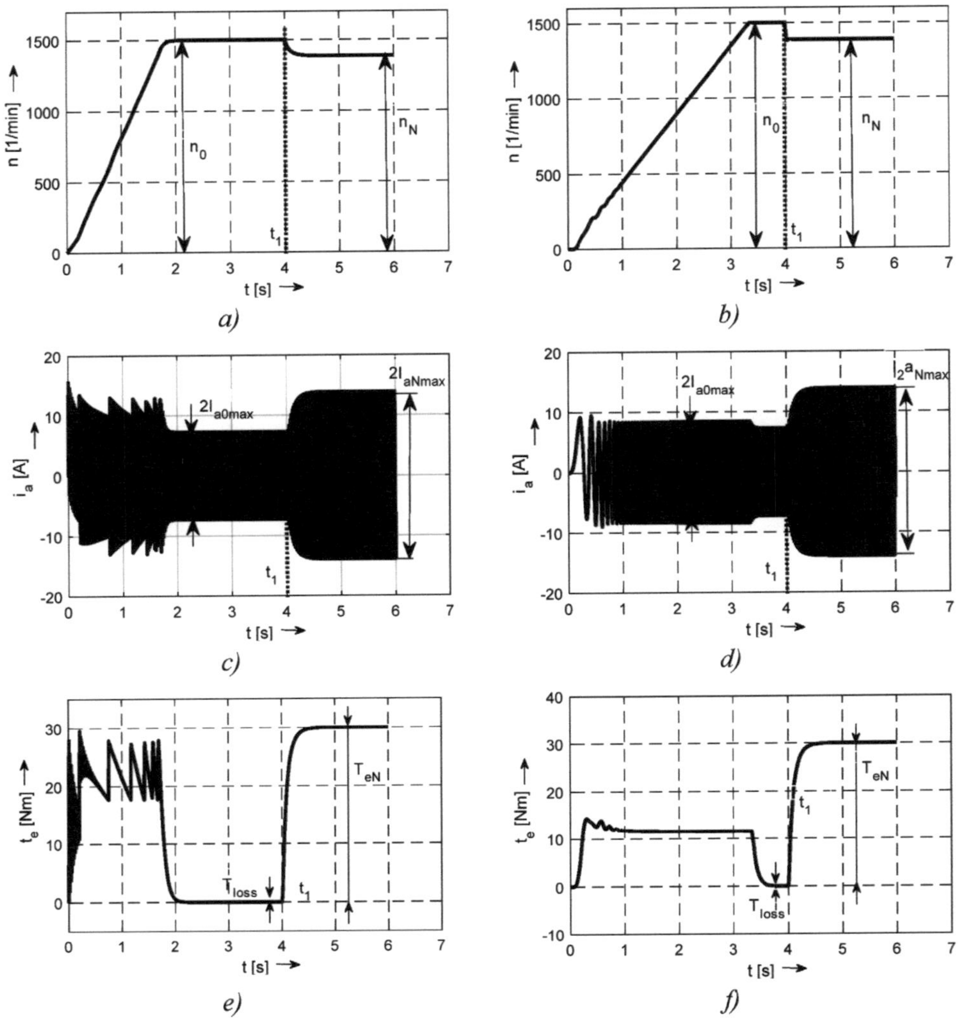

Figure 33.
Time waveforms of the simulations during the starting up of the wound rotor asynchronous motor by means of rheostats added to the rotor circuits: (a) speed n*, (c) phase current* I_a*, (e) developed electromagnetic torque, and time waveforms during the starting up by means of frequency converter, again in the same order: (b) speed* n*, (d) phase current* I_a*, and (f) developed electromagnetic torque.*

where j, k = a, b, c, f, D, Q. For a better review, here are all the equations with the sum of all members:

$$\begin{aligned}
\psi_a &= L_{aa}i_a + L_{ab}i_b + L_{ac}i_c + L_{af}i_f + L_{aD}i_D + L_{aQ}i_Q, \\
\psi_b &= L_{ba}i_a + L_{bb}i_b + L_{bc}i_c + L_{bf}i_f + L_{bD}i_D + L_{bQ}i_Q, \\
\psi_c &= L_{ca}i_a + L_{cb}i_b + L_{cc}i_c + L_{cf}i_f + L_{cD}i_D + L_{cQ}i_Q, \\
\psi_f &= L_{fa}i_a + L_{fb}i_b + L_{fc}i_c + L_{ff}i_f + L_{fD}i_D + L_{fQ}i_Q, \\
\psi_D &= L_{Da}i_a + L_{Db}i_b + L_{Dc}i_c + L_{Df}i_f + L_{DD}i_D + L_{DQ}i_Q, \\
\psi_Q &= L_{Qa}i_a + L_{Qb}i_b + L_{Qc}i_c + L_{Qf}i_f + L_{QD}i_D + L_{QQ}i_Q.
\end{aligned} \tag{321}$$

Although it is known that mutual inductances of the windings that are perpendicular to each other are zero:

$$L_{fQ} = L_{Qf} = L_{DQ} = L_{QD} = 0, \tag{322}$$

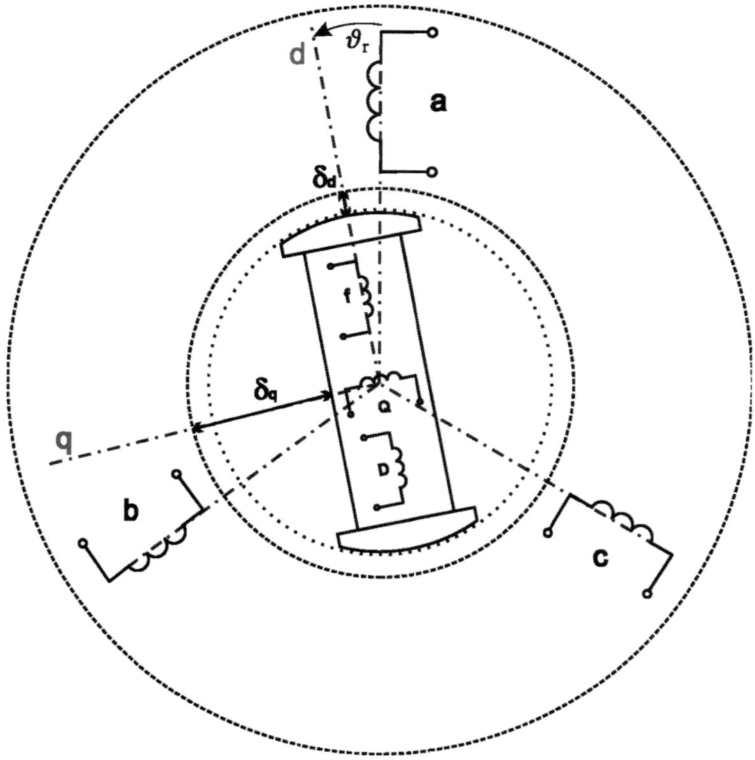

Figure 34.
Synchronous machine with salient poles on the rotor and three-phase winding a, b, c on the stator, field winding f, and damping winding split into two parts (D and Q) perpendicular to each other, positioned in the d-axis and q-axis on the rotor. The d-axis on the rotor is shifted about the angle ϑr from the axis of a-phase on the stator.

for computer manipulation is more suitable if the original structure is kept and all inductances appear during the analysis:

$$L_{aa}, L_{bb}, L_{cc} \text{ are self} - \text{inductances of the stator windings.} \tag{323}$$

$$L_{ff}, L_{DD}, L_{QQ} \text{ are self} - \text{inductances of the rotor windings.} \tag{324}$$

$$L_{ab}, L_{ac}, L_{ba}, L_{bc}, L_{ca}, L_{cb} \text{ are mutual inductances of the stator windings.} \tag{325}$$

$$L_{fD}, L_{fQ}, L_{Df}, L_{DQ}, L_{Qf}, L_{QD} \text{ are mutual inductances of the rotor windings.} \tag{326}$$

The rest of the inductances are mutual inductances of the stator and rotor windings:

$$L_{af}, L_{bf}, L_{cf}, L_{aD}, \text{etc.}$$

It is important to investigate if inductances depend on the rotor position or not.

2.16.1 Inductances that do not depend on the rotor position

Self- and mutual inductances of the rotor windings $L_{ff}, L_{QQ}, L_{DD}, L_{fD}$ do not depend on the rotor position because the stator is cylindrical, and if the stator slotting is neglected, then the air gap is for each winding constant. Thus, the magnetic permeance of the path of magnetic flux created by these windings does not change if the rotor rotates.

2.16.2 Inductances depending on the rotor position

2.16.2.1 Mutual inductances of the rotor and stator windings

Investigate, for example, a-phase winding on the stator and field winding f on the rotor, as it is shown in **Figure 34**.

When sinusoidally distributed windings are assumed, i.e., coefficients of higher harmonic components are zero, then the waveform of mutual inductance is cosinusoidal, if for the origin of the system such rotor position is chosen in which the a-phase axis and the axis of the field winding are identical (see **Figure 35**).

Then the mutual inductances can be expressed as follows:

$$L_{af} = L_{fa} = L_{afmax} \cos \vartheta_r \tag{327}$$

$$L_{bf} = L_{fb} = L_{afmax} \cos\left(\vartheta_r - \frac{2\pi}{3}\right) \tag{328}$$

$$L_{cf} = L_{fc} = L_{afmax} \cos\left(\vartheta_r + \frac{2\pi}{3}\right) \tag{329}$$

similarly:

$$L_{aD} = L_{Da} = L_{aDmax} \cos \vartheta_r \tag{330}$$

$$L_{bD} = L_{Db} = L_{aDmax} \cos\left(\vartheta_r - \frac{2\pi}{3}\right) \tag{331}$$

$$L_{cD} = L_{Dc} = L_{aDmax} \cos\left(\vartheta_r + \frac{2\pi}{3}\right) \tag{332}$$

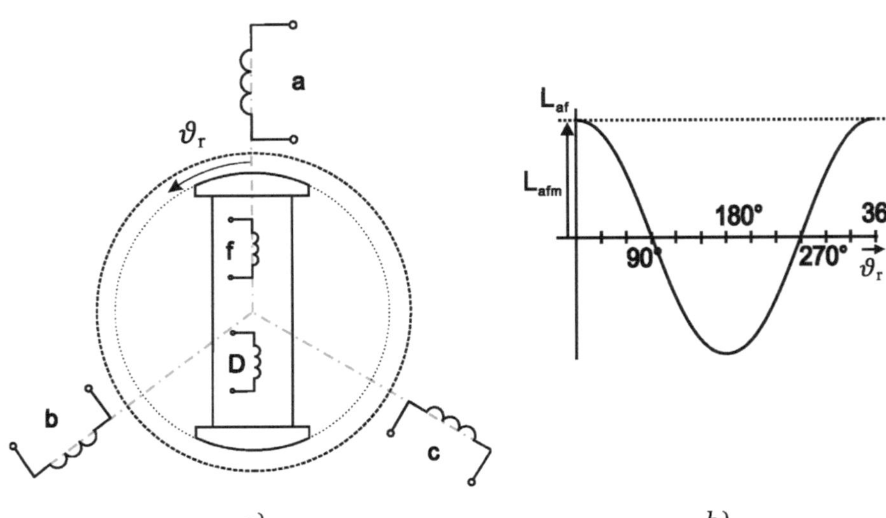

Figure 35.
(a) Illustration to express mutual inductance of the a-phase on the stator and field winding f on the rotor, (b) waveform of the mutual inductance L_{af} versus rotor position ϑ_r.

Expressions for Q-winding positioned in the q-axis are written according to **Figure 36a**, where it is seen that the positive q-axis is ahead about 90° of the d-axis. Hence if the d-axis is identified with the axis of the a-phase, the q-axis is perpendicular to it, and mutual inductance L_{aQ} is zero. To obtain a position in which L_{aQ} is maximal, it is necessary to go back about 90°, to identify q-axis with the a-phase axis. There the L_{aQ} receives its magnitude. The magnitudes of the mutual inductances between Q-winding and b- and c-phases are shifted about 120°, as it is seen in **Figure 36b**.

$$L_{aQ} = L_{Qa} = L_{aQmax}\cos\left(\vartheta_r + \frac{\pi}{2}\right) = -L_{aQmax}\sin\vartheta_r, \tag{333}$$

$$L_{bQ} = L_{Qb} = bL_{aQmax}\sin\left(\vartheta_r - \frac{2\pi}{3}\right), \tag{334}$$

$$L_{cQ} = L_{Qc} = cL_{aQmax}\sin\left(\vartheta_r + \frac{2\pi}{3}\right) \tag{335}$$

2.16.2.2 Self-inductances of the stator

Self-inductances of the stator depend on the rotor position if there are salient poles. Self-inductance of the a-phase is maximal (L_{aamax}), if its axis is identical with the axis of the pole. In this position the magnetic permeance is maximal. The minimal self-inductance of the a-phase (L_{aamin}) occurs if the axis of the a-phase and axis of the pole are shifted about $\pi/2$. Because the magnetic permeance is periodically changed for each pole, it means north and south, the cycle of the self inductance is π, as it is seen in **Figure 37**.

$$L_{aa} = L_{a0} + L_2\cos 2\vartheta_r, \tag{336}$$

$$L_{bb} = L_{a0} + L_2\cos 2\left(\vartheta_r - \frac{2\pi}{3}\right), \tag{337}$$

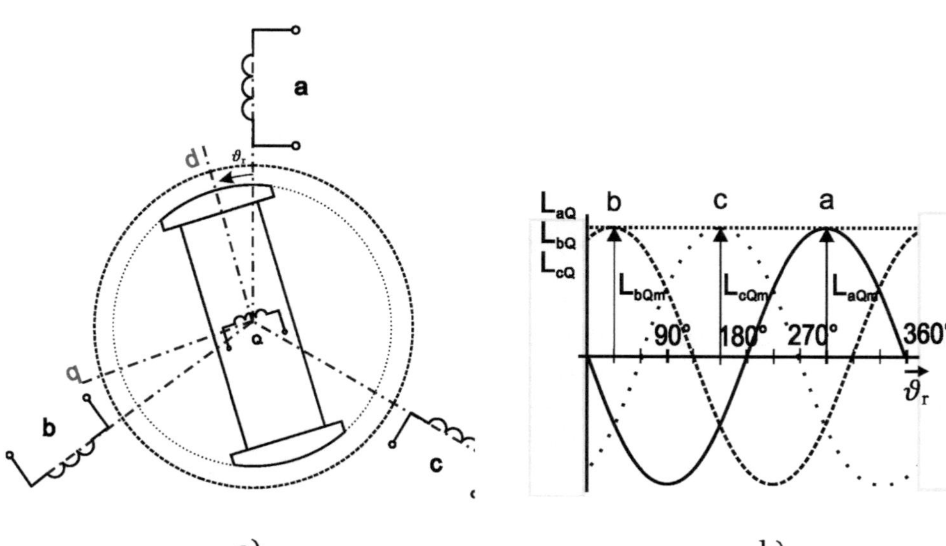

Figure 36.
(a) Illustration to express mutual inductance of the Q-winding on the rotor and a-phase on the stator and (b) waveform of the mutual inductances L_{aQ}, L_{bQ}, L_{cQ}, versus rotor position ϑ_r.

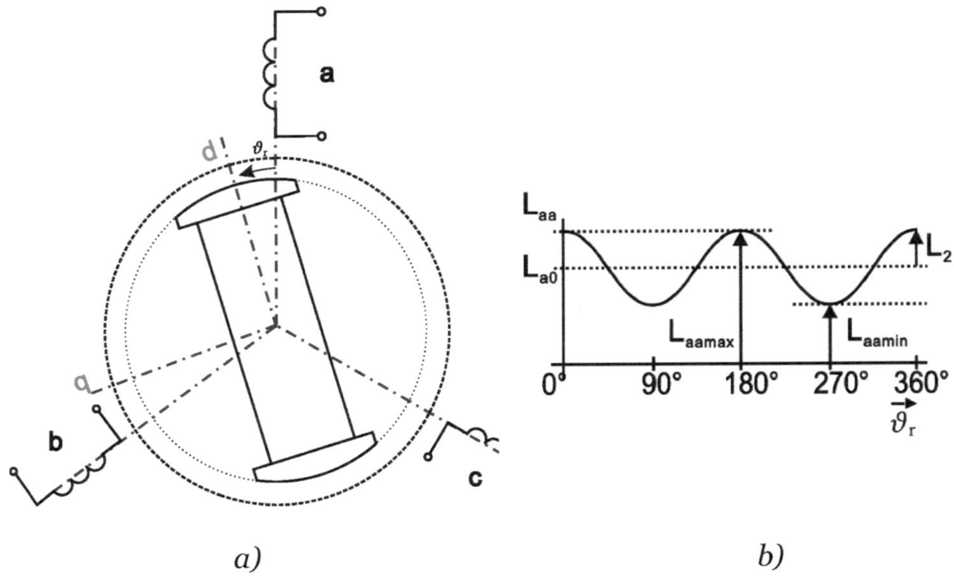

Figure 37.
(a) Illustration to express self-inductance L_{aa} of the a-phase on the stator and (b) waveform of the self-inductance L_{aa} versus rotor position ϑ_r.

$$L_{cc} = L_{a0} + L_2 \cos 2\left(\vartheta_r + \frac{2\pi}{3}\right). \tag{338}$$

The magnitude of the self-inductance L_{aamax} is obtained if the axis of the salient pole is identical with the axis of the stator a-phase; it means $\vartheta_r = 0$. Then:

$$L_{aamax} = L_{a0} + L_2. \tag{339}$$

The minimal value of the self-inductance is obtained if the axis of the salient pole is perpendicular to the axis of the stator a-phase, i.e., $\vartheta_r = \pi/2$. Then:

$$L_{aamin} = L_{a0} - L_2. \tag{340}$$

If the rotor rotates about $\vartheta_r = \pi$, the self-inductance obtains again its maximal value, etc.; accordingly self-inductance does not obtain negative values, as it is seen in **Figure 37b**.

2.16.2.3 Mutual inductance of the stator windings

Mutual inductances of the stator windings depend on the rotor position only in the case of the salient poles on the rotor. These inductances are negative because they are shifted about 120° (see explanation in **Figure 27b**). The rotor is in a position where mutual inductance L_{bc} is maximal is shown in **Figure 38a**. Its waveform vs. rotor position is in **Figure 38b**.

It is possible to assume that for the sinusoidally distributed windings, the magnitudes of harmonic waveform L_2 are the same as in the case of the self-inductance of the stator windings. In the windings embedded in the slots, with a final number of the slots around the rotor periphery and the same number of the conductors in the slots, this assumption is not fulfilled; thus magnitudes of self and mutual waveforms can be different. Here a source of mistakes can be found and eventually

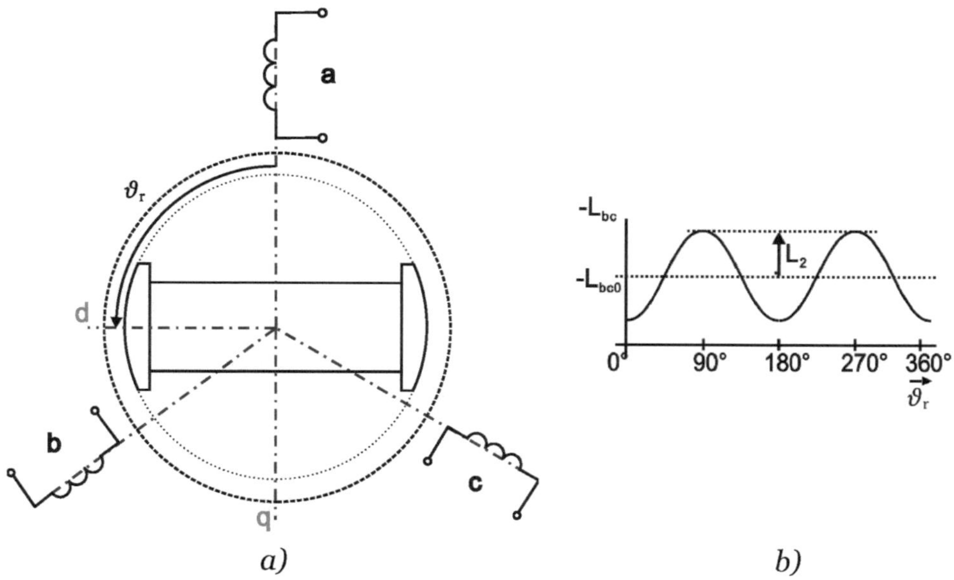

Figure 38.
(a) Illustration to express mutual inductance of the stator windings a, b, c and (b) waveform of the mutual inductance L_{bc} *vs. rotor position* ϑ_r.

discrepancies between the calculated and measured values. The waveforms in **Figure 38b** can be written as follows:

$$-L_{bc} = L_{ab0} - L_2 \cos 2\vartheta_r \tag{341}$$

$$-L_{ca} = L_{ab0} - L_2 \cos 2\left(\vartheta_r - \frac{2\pi}{3}\right) = L_{ab0} - L_2 \cos\left(2\vartheta_r + \frac{2\pi}{3}\right) \tag{342}$$

$$-L_{ab} = L_{ab0} - L_2 \cos 2\left(\vartheta_r + \frac{2\pi}{3}\right) = L_{ab0} - L_2 \cos\left(2\vartheta_r - \frac{2\pi}{3}\right) \tag{343}$$

or

$$L_{ab} = -L_{ab0} + L_2 \cos 2\left(\vartheta_r + \frac{2\pi}{3}\right) = -L_{ab0} + L_2 \cos\left(2\vartheta_r - \frac{2\pi}{3}\right) \tag{344}$$

which better corresponds to the waveform in **Figure 38**.

Now all the expressions of these inductances are introduced into Eq. (**65**) and Eq. (318). They are equations with nonlinear periodically changed coefficients. To eliminate these coefficients, it is necessary to transform the currents, voltages, and linkage magnetic fluxes. The most suitable is Park linear transformation, which was explained in Section 4 and is applied again in the next chapter.

2.17 Terminal voltage equations of the synchronous machine after a transformation into k-system with the axes d, q, 0

Terminal voltage equations of the synchronous machine stator windings in a system d, q, 0 are obtained by means of the procedure described in Section 7. The next equations were derived:

$$u_{\mathrm{d}} = R_{\mathrm{s}} i_{\mathrm{d}} + \frac{\mathrm{d}\psi_{\mathrm{d}}}{\mathrm{d}t} - \omega_{\mathrm{k}} \psi_{\mathrm{q}}, \tag{345}$$

$$u_{\mathrm{q}} = R_{\mathrm{s}} i_{\mathrm{q}} + \frac{\mathrm{d}\psi_{\mathrm{q}}}{\mathrm{d}t} + \omega_{\mathrm{k}} \psi_{\mathrm{d}}, \tag{346}$$

$$u_0 = R_{\mathrm{s}} i_0 + \frac{\mathrm{d}\psi_0}{dt}. \tag{347}$$

Equations (160)–(162) are voltage equations of the three-phase stator windings, in this case synchronous machine but also asynchronous machine, as it was mentioned in Section 13. They are equations transformed into reference k-system rotating by angular speed ω_{k}, with the axes d, q, 0. As it is seen, they are the same equations as in Section 2.1, which were derived for universal configuration of an electrical machine.

Terminal voltage equations of the synchronous machine rotor windings are not needed to transform in the d-axis and q-axis, because the rotor windings are embedded in these axes, as it is seen in **Figure 34**, and are written directly in the two-axis system d, q, 0:

$$u_{\mathrm{f}} = R_{\mathrm{f}} i_{\mathrm{f}} + \frac{\mathrm{d}\psi_{\mathrm{f}}}{\mathrm{d}t}, \tag{348}$$

$$u_{\mathrm{D}} = R_{\mathrm{D}} i_{\mathrm{D}} + \frac{\mathrm{d}\psi_{\mathrm{D}}}{\mathrm{d}t}, \tag{349}$$

$$u_{\mathrm{Q}} = R_{\mathrm{Q}} i_{\mathrm{Q}} + \frac{\mathrm{d}\psi_{\mathrm{Q}}}{\mathrm{d}t}. \tag{350}$$

The next the expressions for linkage magnetic flux are investigated.

2.18 Linkage magnetic flux equations of the synchronous machine in the general theory of electrical machines

In Eq. (65) of linkage magnetic fluxes, expressions for inductances as they were derived in Section 16 are introduced. For example, for field winding with a subscript "f," the equation for linkage magnetic flux is written as follows:

$$\psi_{\mathrm{f}} = L_{\mathrm{fa}} i_{\mathrm{a}} + L_{\mathrm{fb}} i_{\mathrm{b}} + L_{\mathrm{fc}} i_{\mathrm{c}} + L_{\mathrm{ff}} i_{\mathrm{f}} + L_{\mathrm{fD}} i_{\mathrm{D}} + L_{\mathrm{fQ}} i_{\mathrm{Q}}, \tag{351}$$

$$\psi_{\mathrm{f}} = L_{\mathrm{afmax}} \left[i_{\mathrm{a}} \cos \vartheta_{\mathrm{r}} + i_{\mathrm{b}} \cos\left(\vartheta_{\mathrm{r}} - \frac{2\pi}{3}\right) + i_{\mathrm{c}} \cos\left(\vartheta_{\mathrm{r}} + \frac{2\pi}{3}\right)\right] + L_{\mathrm{ff}} i_{\mathrm{f}} + L_{\mathrm{fD}} i_{\mathrm{D}} + L_{\mathrm{fQ}} i_{\mathrm{Q}}. \tag{352}$$

If this equation is compared with Eq. (124), written for a general variable x, it is seen that the expression in the square bracket is equal to $i_{\mathrm{d}}/k_{\mathrm{d}}$, if $\vartheta_{\mathrm{k}} = \vartheta_{\mathrm{r}}$:

$$\left[i_{\mathrm{a}} \cos \vartheta_{\mathrm{r}} + i_{\mathrm{b}} \cos\left(\vartheta_{\mathrm{r}} - \frac{2\pi}{3}\right) + i_{\mathrm{c}} \cos\left(\vartheta_{\mathrm{r}} + \frac{2\pi}{3}\right)\right] = \frac{1}{k_{\mathrm{d}}} i_{\mathrm{d}}. \tag{353}$$

After this modification in Eq. (352), the following is obtained:

$$\psi_{\mathrm{f}} = \frac{1}{k_{\mathrm{d}}} L_{\mathrm{afmax}} i_{\mathrm{d}} + L_{\mathrm{ff}} i_{\mathrm{f}} + L_{\mathrm{fD}} i_{\mathrm{D}} + L_{\mathrm{fQ}} i_{\mathrm{Q}} = L_{\mathrm{fd}} i_{\mathrm{d}} + L_{\mathrm{ff}} i_{\mathrm{f}} + L_{\mathrm{fD}} i_{\mathrm{D}}, \tag{354}$$

where it was taken into account that mutual inductance of two perpendicular windings f and Q is zero.

On the same basis, the linkage magnetic flux for damping rotor windings D and Q is received:

$$\psi_{\mathrm{D}} = \frac{1}{k_{\mathrm{d}}} L_{\mathrm{aDmax}} i_{\mathrm{d}} + L_{\mathrm{Df}} i_{\mathrm{f}} + L_{\mathrm{DD}} i_{\mathrm{D}} + L_{\mathrm{DQ}} i_{\mathrm{Q}} = L_{\mathrm{Dd}} i_{\mathrm{d}} + L_{\mathrm{Df}} i_{\mathrm{f}} + L_{\mathrm{DD}} i_{\mathrm{D}}, \tag{355}$$

$$\begin{aligned}\psi_{\mathrm{Q}} &= -L_{\mathrm{aQmax}} \left[i_{\mathrm{a}} \sin \vartheta_{\mathrm{r}} + i_{\mathrm{b}} \sin \left(\vartheta_{\mathrm{r}} - \frac{2\pi}{3} \right) + i_{\mathrm{c}} \sin \left(\vartheta_{\mathrm{r}} + \frac{2\pi}{3} \right) \right] + L_{\mathrm{Qf}} i_{\mathrm{f}} + L_{\mathrm{QD}} i_{\mathrm{D}} \\ &\quad + L_{\mathrm{QQ}} i_{\mathrm{Q}} \\ &= \frac{1}{k_{\mathrm{q}}} L_{\mathrm{aQmax}} i_{\mathrm{q}} + L_{\mathrm{Qf}} i_{\mathrm{f}} + L_{\mathrm{QD}} i_{\mathrm{D}} + L_{\mathrm{QQ}} i_{\mathrm{Q}} = L_{\mathrm{Qq}} i_{\mathrm{q}} + L_{\mathrm{QQ}} i_{\mathrm{Q}}\end{aligned} \tag{356}$$

A derivation for the stator windings is made in the like manner. It is started with a formal transformation equation from system a, b, c into the d-axis and then into the q-axis. The equation in the d-axis is as follows:

$$\psi_{\mathrm{d}} = k_{\mathrm{d}} \left[\psi_{\mathrm{a}} \cos \vartheta_{\mathrm{r}} + \psi_{\mathrm{b}} \cos \left(\vartheta_{\mathrm{r}} - \frac{2}{3}\pi \right) + \psi_{\mathrm{c}} \cos \left(\vartheta_{\mathrm{r}} + \frac{2}{3}\pi \right) \right]. \tag{357}$$

If into this equation expressions from Eq. (321), for linkage magnetic fluxes of a, b, c phases, are introduced, and for inductances appropriate expressions from Section 16 are introduced, then after widespread modifications of the goniometrical functions and for a rotor position in d-axis, i.e., if

$$\vartheta_{\mathrm{k}} = \vartheta_{\mathrm{r}} = 0,$$

the following is received:

$$\psi_{\mathrm{d}} = L_{\mathrm{d}} i_{\mathrm{d}} + \frac{3}{2} k_{\mathrm{d}} L_{\mathrm{afmax}} i_{\mathrm{f}} + \frac{3}{2} k_{\mathrm{d}} L_{\mathrm{aDmax}} i_{\mathrm{D}} = L_{\mathrm{d}} i_{\mathrm{d}} + L_{\mathrm{df}} i_{\mathrm{f}} + L_{\mathrm{dD}} i_{\mathrm{D}}. \tag{358}$$

Here:

$$L_{\mathrm{d}} = L_{\mathrm{a0}} + L_{\mathrm{ab0}} + \frac{3}{2} L_2 \tag{359}$$

is a direct synchronous inductance. The other symbols are for mutual inductances between the stator windings transformed into the d-axis and rotor windings, which are also in the d-axis:

$$L_{\mathrm{df}} = \frac{3}{2} k_{\mathrm{d}} L_{\mathrm{afmax}}, \tag{360}$$

$$L_{\mathrm{dD}} = \frac{3}{2} k_{\mathrm{d}} L_{\mathrm{aDmax}}. \tag{361}$$

The linkage magnetic flux in the q-axis is derived in a similar way, which results in:

$$\psi_{\mathrm{q}} = L_{\mathrm{q}} i_{\mathrm{q}} + \frac{3}{2} k_{\mathrm{q}} L_{\mathrm{aQmax}} i_{\mathrm{Q}} = L_{\mathrm{q}} i_{\mathrm{q}} + L_{\mathrm{qQ}} i_{\mathrm{Q}}, \tag{362}$$

where:

$$L_q = L_{a0} + L_{ab0} - \frac{3}{2}L_2 \tag{363}$$

is a quadrature synchronous inductance and

$$L_{qQ} = \frac{3}{2}k_q L_{aQmax}, \tag{364}$$

is the mutual inductance between the stator winding transformed into the q-axis and rotor winding which is in the q-axis . From Eqs. (359) and (363), it is seen that if there is a cylindrical rotor, then $L_2 = 0$, and inductances in both axes are equal:

$$L_d = L_q, \tag{365}$$

which is a known fact.

Finally, a linkage magnetic flux for zero axis is derived in a similar way:

$$\psi_0 = L_0 i_0, \tag{366}$$

where:

$$L_0 = L_{a0} - 2L_{ab0} \tag{367}$$

is called zero inductance. It is seen that this linkage magnetic flux and inductance are linked only with variables with the subscript 0 and do not have any relation to the variables in the other axes. Additionally, also here a knowledge from the theory of the asynchronous machine can be applied that zero impedance is equal to the leakage stator inductance that can be used during the measurement of the leakage stator inductance. Here can be reminded equation Section 12:

$$L_0 = L_s - 2M_s = L_{\sigma s} + M - 2\frac{M}{2} = L_{\sigma s}. \tag{368}$$

Namely, if all three phases of the stator windings (in series or parallel connection) are fed by a single-phase voltage, it results in the pulse, non-rotating magnetic flux (see [8]).

If there is a request to make equations more simple, then it is necessary to ask for equality of mutual inductances of two windings, for example, inductance L_{fd} for the current i_d should be equal to the inductance L_{df} for the current i_f:

$$L_{fd} = L_{df}. \tag{369}$$

Therefore from Eq. (354) for ψ_f, take the expression at the current i_d, which was marked as L_{fd} and put into the equality with the expression at the current i_f in Eq. (358) for ψ_d, which was marked as L_{df}:

$$\frac{1}{k_d}L_{afmax} = \frac{3}{2}k_d L_{afmax}. \tag{370}$$

Then:

$$k_d^2 = \frac{2}{3}, \tag{371}$$

and

$$k_{\mathrm{d}} = \sqrt{\frac{2}{3}}. \tag{372}$$

The same value is obtained if expressions for ψ_{d} at the current i_{D} (358) and ψ_{D} at the current i_{d} (355) are put into the equality, to get $L_{\mathrm{Dd}} = L_{\mathrm{dD}}$. Then:

$$\frac{1}{k_{\mathrm{d}}} L_{\mathrm{aDmax}} = \frac{3}{2} k_{\mathrm{d}} L_{\mathrm{aDmax}}, \tag{373}$$

which results in $k_{\mathrm{d}} = \sqrt{\frac{2}{3}}$.

In the q-axis it is done at the same approach: The expression at the current i_{q} in the equation for ψ_{Q} (356) and the expression at the current i_{Q} in equation for ψ_{q} (362), put into equality to get $L_{\mathrm{Qq}} = L_{\mathrm{qQ}}$, are as follows:

$$\frac{1}{k_q} L_{\mathrm{aQmax}} = \frac{3}{2} k_q L_{\mathrm{aQ}\,max}\,. \tag{374}$$

It results in the value:

$$k_{\mathrm{q}} = \sqrt{\frac{2}{3}}. \tag{375}$$

Hence a choice for the coefficients suitable for synchronous machines flows:

$$k_{\mathrm{d}} = k_{\mathrm{q}} = \pm\sqrt{\frac{2}{3}},$$

but it is better to use the positive expression:

$$k_{\mathrm{d}} = k_{\mathrm{q}} = \sqrt{\frac{2}{3}}. \tag{376}$$

If the next expressions are introduced:

$$L_{\mathrm{df}} = \frac{3}{2}\sqrt{\frac{2}{3}} L_{\mathrm{afmax}} = \sqrt{\frac{3}{2}} L_{\mathrm{afmax}} = L_{\mathrm{fd}}, \tag{377}$$

$$L_{\mathrm{dD}} = \frac{3}{2}\sqrt{\frac{2}{3}} L_{\mathrm{aDmax}} = \sqrt{\frac{3}{2}} L_{\mathrm{aDmax}} = L_{\mathrm{Dd}}, \tag{378}$$

$$L_{\mathrm{qQ}} = \frac{3}{2}\sqrt{\frac{2}{3}} L_{\mathrm{aQmax}} = \sqrt{\frac{3}{2}} L_{\mathrm{aQmax}} = L_{\mathrm{Qq}}, \tag{379}$$

then equations for linkage magnetic fluxes of the synchronous machines in the d, q, 0 system have the form as follows:

$$\begin{aligned} \psi_{\mathrm{d}} &= L_{\mathrm{d}} i_{\mathrm{d}} + L_{\mathrm{df}} i_{\mathrm{f}} + L_{\mathrm{dD}} i_{\mathrm{D}}, \text{see } (358) \\ \psi_{\mathrm{q}} &= L_{\mathrm{q}} i_{\mathrm{q}} + L_{\mathrm{qQ}} i_{\mathrm{Q}}, \text{see } (362) \\ \psi_0 &= L_0 i_0, \text{see } (366) \end{aligned} \tag{380}$$

$$\psi_{\mathrm{f}} = L_{\mathrm{fd}} i_{\mathrm{d}} + L_{\mathrm{ff}} i_{\mathrm{f}} + L_{\mathrm{fD}} i_{\mathrm{D}}, \text{see } (354)$$
$$\psi_{\mathrm{D}} = L_{\mathrm{Dd}} i_{\mathrm{d}} + L_{\mathrm{Df}} i_{\mathrm{f}} + L_{\mathrm{DD}} i_{\mathrm{D}}, \text{see } (355)$$
$$\psi_{\mathrm{Q}} = L_{\mathrm{Qq}} i_{\mathrm{q}} + L_{\mathrm{QQ}} i_{\mathrm{Q}}, \text{see } (356)$$

where:

$$\begin{aligned} L_{\mathrm{d}} &= L_{\mathrm{a0}} + L_{\mathrm{ab0}} + \frac{3}{2} L_2, \text{see } (359) \\ L_{\mathrm{q}} &= L_{\mathrm{a0}} + L_{\mathrm{ab0}} - \frac{3}{2} L_2, \text{see } (363) \\ L_0 &= L_{\mathrm{a0}} - 2L_{\mathrm{ab0}}, \text{see } (367) \\ L_{\mathrm{df}} &= L_{\mathrm{fd}} = \sqrt{\frac{3}{2}} L_{\mathrm{afmax}}, \text{see } (377) \\ L_{\mathrm{dD}} &= L_{\mathrm{Dd}} = \sqrt{\frac{3}{2}} L_{\mathrm{aDmax}}, \text{see } (378) \\ L_{\mathrm{qQ}} &= L_{\mathrm{Qq}} = \sqrt{\frac{3}{2}} L_{\mathrm{aQmax}}, \text{see } (379). \end{aligned} \tag{381}$$

By this record it was proven that not only terminal voltage equations of the stator and rotor windings but also equations of the linkage magnetic fluxes are identical with the equations of the universal electrical machine. Of course, expressions for inductances and a mode of their measurements are changed according the concrete electrical machine.

To complete a system of equations, it is necessary to add the equation for angular speed and to derive expression for the electromagnetic torque.

2.19 Power and electromagnetic torque of the synchronous machine

The instantaneous value of electrical input power in the a, b, c system can be written in the same form as it was derived in Section 8:

$$p_{\mathrm{in}} = u_{\mathrm{a}} i_{\mathrm{a}} + u_{\mathrm{b}} i_{\mathrm{b}} + u_{\mathrm{c}} i_{\mathrm{c}}$$

and also in the d, q, 0 system:

$$p_{\mathrm{in}} = \frac{2}{3} \frac{1}{k_{\mathrm{d}}^2} u_{\mathrm{d}} i_{\mathrm{d}} + \frac{2}{3} \frac{1}{k_{\mathrm{q}}^2} u_{\mathrm{q}} i_{\mathrm{q}} + \frac{1}{3} \frac{1}{k_0^2} u_0 i_0.$$

If the constants recommended in the previous chapter are employed,

$$k_{\mathrm{d}} = k_{\mathrm{q}} = \sqrt{\frac{2}{3}} k_0 = \sqrt{\frac{1}{3}}$$

The power of synchronous machine is obtained in the form:

$$p_{\mathrm{in}} = u_{\mathrm{d}} i_{\mathrm{d}} + u_{\mathrm{q}} i_{\mathrm{q}} + u_0 i_0,$$

for which the principle of the power invariancy is valid.

An expression for the instantaneous value of electromagnetic torque of synchronous machine is derived from the energy balance of the machine. If the stator variables are transformed and the rotor variables remained in their real form, on the basis of Eq. (184), the following can be written:

$$p_{\mathrm{in}} = \frac{2}{3}\frac{1}{k_{\mathrm{d}}^2}u_{\mathrm{d}}i_{\mathrm{d}} + \frac{2}{3}\frac{1}{k_{\mathrm{q}}^2}u_{\mathrm{q}}i_{\mathrm{q}} + \frac{1}{3}\frac{1}{k_0^2}u_0 i_0 + u_{\mathrm{f}}i_{\mathrm{f}} + u_{\mathrm{D}}i_{\mathrm{D}} + u_{\mathrm{Q}}i_{\mathrm{Q}}. \tag{382}$$

This electrical input power for motor operation equals the Joule resistance losses, time varying of magnetic field energy, and internal converted power from the electrical to the mechanical form. If this internal converted power is given only by rotating members of the voltages, which are seen in the equations for u_{d} and u_{q}, then after introduction and modification, the same equation as in (173) is received:

$$t_{\mathrm{e}} = p\frac{2}{3}\frac{1}{k_{\mathrm{d}}k_{\mathrm{q}}}\left(\psi_{\mathrm{d}}i_{\mathrm{q}} - \psi_{\mathrm{q}}i_{\mathrm{d}}\right).$$

If the constants as they were derived above are introduced, the next expression for the electromagnetic torque is obtained:

$$t_{\mathrm{e}} = p\left(\psi_{\mathrm{d}}i_{\mathrm{q}} - \psi_{\mathrm{q}}i_{\mathrm{d}}\right).$$

This expression is identical with that derived for universal electrical machine in Section 2, Eq. (82).

In the next chapters, here derived equations will be applied on a concrete synchronous machine, and transients will be investigated. Again, this reminds that all rotor variables are referred to the stator.

2.20 Synchronous machine in the dq0 system

In Section 17 there are derived terminal voltage equations of the synchronous machine stator windings, transformed into k-reference frame, rotating by angular speed ω_{k}, with the d, q, 0 axis. It was not needed to transform the rotor voltage equations, because the rotor windings are really embedded in the d-axis and q-axis, as it is seen in **Figure 34** and is written directly in the d, q, 0 system.

Then it looks suitable to investigate transients of synchronous machine in the system d, q, 0. It means that it is necessary to identify the reference k-system with the rotor windings in such a way that the d-axis is identical with field winding axis and q-axis, which is perpendicular to it, as it is known from the arrangement of the classical synchronous machine with salient poles.

This reminds that an analysis of the synchronous machine armature reaction, with a splitting into two perpendicular d-axis and q-axis, is the basis for the general theory of the electrical machine. Therefore, it is justified to mark the subscripts d, q, according the axes. Then the k-system rotates by the rotor speed, and the angle between the positive real axis of the k-system and stator a-phase axis is identical with the angle of the rotor d-axis:

$$\omega_{\mathrm{k}} = \omega_{\mathrm{r}}$$

$$\vartheta_{\mathrm{k}} = \vartheta_{\mathrm{r}}. \tag{383}$$

Respecting these facts the stator voltage equations from Section 17 are in the form:

$$u_d = R_s i_d + \frac{d\psi_d}{dt} - \omega_r \psi_q, \tag{384}$$

$$u_q = R_s i_q + \frac{d\psi_q}{dt} + \omega_r \psi_d, \tag{385}$$

$$u_0 = R_s i_0 + \frac{d\psi_0}{dt}. \tag{386}$$

The rotor equations are without any change:

$$u_f = R_f i_f + \frac{d\psi_f}{dt}, \tag{387}$$

$$u_D = R_D i_D + \frac{d\psi_D}{dt}, \tag{388}$$

$$u_Q = R_Q i_Q + \frac{d\psi_Q}{dt}. \tag{389}$$

Linkage magnetic fluxes and inductances were given in Equations (380) and (381).

2.20.1 Relation to the parameters of the classical equivalent circuit

Now is the time to explain a relation between the terminology and inductance marking of the classical theory of synchronous machine and its equivalent circuit and general theory of electrical machines. As it is known, a term "reactance of the armature reaction," and also inductance of the armature reaction, corresponds to the magnetizing reactance (inductance). The sum of the magnetizing reactance $X_{\mu d}$ (inductance $L_{\mu d}$) and leakage reactance $X_{\sigma s}$ (inductance $L_{\sigma s}$) creates synchronous reactance (inductance) in the relevant axis; thus for the synchronous reactance (inductance) in the d-axis, the following can be written:

$$X_d = X_{ad} + X_{\sigma s} = X_{\mu d} + X_{\sigma s},$$
$$L_d = L_{ad} + L_{\sigma s} = L_{\mu d} + L_{\sigma s}, \tag{390}$$

and for the synchronous reactance (inductance) in the q-axis, the following can be written:

$$X_q = X_{aq} + X_{\sigma s} = X_{\mu q} + X_{\sigma s},$$
$$L_q = L_{aq} + L_{\sigma s} = L_{\mu q} + L_{\sigma s}. \tag{391}$$

For the rotor windings, it is valid that the self-inductance of the field winding L_{ff} is a sum of a mutual inductance of the windings on stator side in the d-axis (because the field winding is also in d-axis) $L_{\mu d}$ and leakage inductance of the field winding $L_{\sigma f}$. Both windings must be with the same number of turns; in other words both windings must be on the same side of the machine. It is suitable to refer the rotor windings to the stator side and mark them with the "′(prime)," as it is known from the theory of transformer, where secondary variables are referred to the primary side, or rotor variables of the asynchronous machine referred to the stator side. The mutual (magnetizing) inductance in the d-axis is defined and measured from the stator side; therefore, it is not necessary to refer it:

$$L'_{\mathrm{ff}} = L'_{\sigma\mathrm{f}} + L_{\mu\mathrm{d}}. \tag{392}$$

The same principle is applied also for the damping windings in both axes. For damping winding in the d-axis, the following is valid:

$$L'_{\mathrm{DD}} = L'_{\sigma\mathrm{D}} + L_{\mu\mathrm{d}}, \tag{393}$$

and for damping winding in q-axis:

$$L'_{\mathrm{QQ}} = L'_{\sigma\mathrm{Q}} + L_{\mu\mathrm{q}}. \tag{394}$$

A factor by which the rotor variables are referred to the stator is $\frac{3}{2}g^2$, where g is a factor needed to refer variables between the stator and rotor side, known from the classical theory of the synchronous machine. Its second power can be justified by the theory of the impedance referring or inductances in transformer or asynchronous machine theory. The constant 3/2 is there because of the referring between three-phase and two-axis systems. Then the leakage inductance of the field winding referred to the stator side is:

$$L'_{\sigma\mathrm{f}} = \frac{3}{2} g^2 L_{\sigma\mathrm{f}}. \tag{395}$$

For a more detailed explanation, it is useful to mention that the factor g is defined for referring the stator current to the rotor current, e.g., the stator armature current I_a referred to the rotor side is $I_\mathrm{a}' = gI_\mathrm{a}$, which is needed for phasor diagrams.

As it is known from the theory of transformers and asynchronous machines, the current ratio is the inverse value of the voltage ratio and eventually of the ratio of the number of turns. If the impedance of rotor is referred to the stator, it is made by the second power of the voltage ratio (or number of turns ratio) of the stator and the rotor side. Because in the synchronous machine, a current coefficient (subscript I) from the stator to the rotor (subscript sr) is obviously applied and marked with g, it will be shown that the voltage ratio (subscript U) is its inverse value:

$$g_{\mathrm{Isr}} = \frac{I_{\mathrm{fo}}}{I_{\mathrm{k0}}} = \frac{1}{g_{U\mathrm{sr}}} = \frac{1}{\frac{1}{I_{\mathrm{f0}}/I_{\mathrm{k0}}}} = \frac{1}{\frac{I}{I_{\mathrm{f0}}}}$$

$$g_{\mathrm{Usr}} = \frac{I_{\mathrm{k0}}}{I_{\mathrm{f0}}}. \tag{396}$$

Shortly to explain, the I_{f0} is the field current at which the rated voltage is induced in the stator winding at no load condition, and I_{k0} is the stator armature current which flows in the stator winding, if the terminals are applied to the rated voltage at zero excitation (zero field current), or in other words, it is the current which flows in the stator winding at short circuit test, if the field current is I_{f0}.

If rotor variables should be referred to the stator (subscript rs; in Eqs. (392)–(395) these variables are marked with ′(prime)), it will be once more inverse value of the voltage ratio, and hence it is again current ratio g, which is possible to be written shortly as follows:

$$\mathrm{g}_{\mathrm{Urs}} = \frac{I_{\mathrm{f0}}}{I_{\mathrm{k0}}} = \mathrm{g}_{\mathrm{Isr}} = \mathrm{g}. \tag{397}$$

Similar to Eq. (395), also the other variables of the rotor windings would be referred to the stator:

$$L'_{\sigma D} = \frac{3}{2} g^2 L_{\sigma D}, \tag{398}$$

$$L'_{\sigma Q} = \frac{3}{2} g^2 L_{\sigma Q}, \tag{399}$$

$$R'_f = \frac{3}{2} g^2 R_f. \tag{400}$$

The mutual inductance of the field winding and windings in the d-axis is in the classical theory of the synchronous machine called magnetizing inductance in the d-axis; also the mutual inductance of the damping winding in the d-axis is the same magnetizing inductance; therefore, it can be expressed (see also Section 12, where the expressions for asynchronous machine are derived):

$$L_{df} = L_{fd} = L_{dD} = L_{Dd} = L_{\mu d}. \tag{401}$$

Likewise it is valid in the q-axis:

$$L_{qQ} = L_{Qq} = L_{\mu q}. \tag{402}$$

Zero inductance was derived in Section 12 for asynchronous machine:

$$L_0 = L_s - 2M_s = L_{\sigma s} + M - 2\frac{M}{2} = L_{\sigma s}. \tag{403}$$

Then it is clear that the zero inductance for synchronous machine is equal to its leakage inductance of the stator winding and is measured in the same manner:

$$L_0 = L_{\sigma s}. \tag{404}$$

Equations for the linkage magnetic fluxes can be rewritten in the form, where the parameters of the synchronous machine known from its classical theory are respected, with a note that all rotor variables are referred to the stator side:

$$\psi_d = L_d i_d + L_{\mu d} i'_f + L_{\mu d} i'_D = (L_{\sigma s} + L_{\mu d}) i_d + L_{\mu d}(i'_f + i'_D), \tag{405}$$

$$\psi_q = L_q i_q + L_{\mu q} i'_Q = (L_{\sigma s} + L_{\mu q}) i_q + L_{\mu q} i'_Q, \tag{406}$$

$$\psi_0 = L_0 i_0 = L_{\sigma s} i_0, \tag{407}$$

$$\psi'_f = L_{\mu d} i_d + (L'_{\sigma f} + L_{\mu d}) i'_f + L_{\mu d} i'_D = L'_{ff} i'_f + L_{\mu d}(i_d + i'_D), \tag{408}$$

$$\psi'_D = L_{\mu d} i_d + L_{\mu d} i'_f + (L'_{\sigma D} + L_{\mu d}) i'_D = L_{\mu d}(i_d + i'_f) + L'_{DD} i'_D \tag{409}$$

$$\psi'_Q = L_{\mu q} i_q + \left(L'_{\sigma Q} + L_{\mu q}\right) i'_Q = L_{\mu q} i_q + L'_{QQ} i'_Q. \tag{410}$$

The rotor terminal voltage equations written for the rotor variables (resistances, linkage magnetic fluxes, terminal voltages) referred to the stator side are as follows:

$$u'_f = R'_f i'_f + \frac{d\psi'_f}{dt}, \tag{411}$$

$$u'_D = R'_D i'_D + \frac{d\psi'_D}{dt}, \tag{412}$$

$$u'_Q = R'_Q i'_Q + \frac{d\psi'_Q}{dt}. \tag{413}$$

In this way, a system of six differential equations is obtained, namely, three for stator windings (384)–(386) and three for rotor windings, the parameters of which are referred to the stator (411)–(413). The linkage magnetic fluxes are written in Eqs. (405)–(410).

2.20.2 Equations for terminal voltages of the stator windings

If the motoring operation is investigated, then it is necessary to derive what the terminal voltage on the left side of Eqs. (384) and (385) means. The terminal voltages are known variables in the motoring operation, but they are sinusoidal variables of the three-phase system, which must be transformed into the system dq0. Therefore it is necessary to go back to Section 9, where the expressions for the transformed sinusoidal variables were derived. If the voltages of the three-phase system were assumed as sinusoidal functions, the voltage in the d-axis was derived in the form:

$$u_d = k_d U_{max} \frac{3}{2} \sin(\omega_s t - \vartheta_k) \tag{414}$$

and in the q-axis in the form:

$$u_q = -k_q U_{max} \frac{3}{2} \cos(\omega_s t - \vartheta_k). \tag{415}$$

The voltage u_0 for the symmetrical three-phase system is zero.

It is seen that concrete expressions for these voltages depend on the k_d, k_q choice and a choice of the reference system position. In Section 18 there were derived expressions, in which the coefficients suitable for synchronous machines were defined as follows:

$$k_d = k_q = \sqrt{\frac{2}{3}}. \tag{416}$$

These coefficients are introduced to the equations for voltages in the d and q axes, and simultaneously it is taken into account that the reference frame dq0 is identified with the rotor, i.e., $\vartheta_k = \vartheta_r$:

$$u_d = \sqrt{\frac{2}{3}} U_{max} \frac{3}{2} \sin(\omega_s t - \vartheta_r) = \sqrt{\frac{3}{2}} U_{max} \sin(\omega_s t - \vartheta_r), \tag{417}$$

and for the q-axis voltage:

$$u_q = -\sqrt{\frac{2}{3}} U_{max} \frac{3}{2} \cos(\omega_s t - \vartheta_r) = -\sqrt{\frac{3}{2}} U_{max} \cos(\omega_s t - \vartheta_r). \tag{418}$$

These expressions can be introduced to the left side of Eqs. (384) and (385).

As it was mentioned in Section 9, for the synchronous machine, it is more suitable to employ voltages of the three-phase system as the cosinusoidal time functions (see (207)–(209)) and for the d-axis and q-axis voltages to use the derived expressions (210) and (211). Together with the mentioned coefficients and after the reference frame is positioned to the rotor, the next expressions could be employed:

$$u_{\mathrm{d}} = \sqrt{\frac{2}{3}} U_{\max} \frac{3}{2} \cos\left(\omega_{\mathrm{s}} t - \vartheta_{\mathrm{r}}\right) = \sqrt{\frac{3}{2}} U_{\max} \cos\left(\omega_{\mathrm{s}} t - \vartheta_{\mathrm{r}}\right), \tag{419}$$

$$u_{\mathrm{q}} = \sqrt{\frac{2}{3}} U_{\max} \frac{3}{2} \sin\left(\omega_{\mathrm{s}} t - \vartheta_{\mathrm{r}}\right) = \sqrt{\frac{3}{2}} U_{\max} \sin\left(\omega_{\mathrm{s}} t - \vartheta_{\mathrm{r}}\right). \tag{420}$$

In **Figure 39a**, there are sinusoidal waveforms of i, B, H, ϕ, and in **Figure 39b**, there is a phase distribution in the slots of the cylindrical stator. The phasor of the resultant magnetic field at the instant t_0, when the a-phase current is zero, is seen in **Figure 39c**, and at the instant t_1, when a-phase current is maximal, it is seen in **Figure 39d**.

In **Figure 39d**, the position and direction of the resultant magnetic field phasor mean magnitude of the rotating magnetic field, identical with a contribution of the a-phase. The load angle, as an angle between the magnitudes of the rotating magnetic field, in this case identical with the a-phase axis and instantaneous rotor position, is measured from this point.

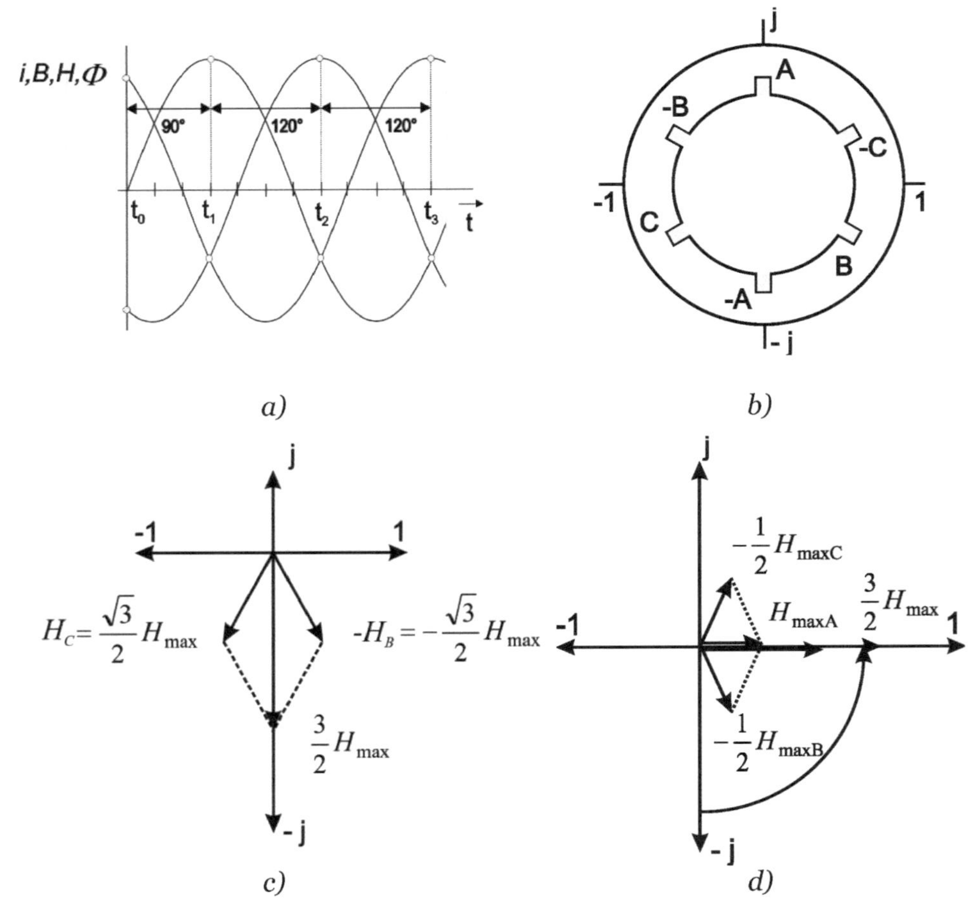

Figure 39.
(a) Sinusoidal waveforms of the variables i, B, H, ϕ *of the symmetrical three-phase system, (b) phase distribution in the slots of the cylindrical stator, (c) phasor of the resultant magnetic field at the instant* t_o*, and (d) phasor of the resultant magnetic field at the instant* t_1*, when the a-phase current is maximal.*

If stator voltage (or current) is investigated from an instant where the current in a-phase is maximal, then the current is described by cosinusoidal function. The other two phases are also cosinusoidal functions, shifted by about 120°. Transformation into the d, q-system results in Eqs. (419) and (420). The load angle is given by the calculation based on Eq. (436).

However, if the origin of the a-phase current waveform is put into zero, it means at instant t_0, the waveform is described by the sinusoidal function, and the phasor of the resultant magnetic field is at the instant t_0, which is the magnitude of the rotating magnetic field, shifted about 90° from the a-phase axis, as it is seen in **Figure 39c**. Since the load angle is investigated from the a-phase axis, it is necessary to subtract that 90° from the calculated value. Then the result will be identical with that gained in the previous case.

2.20.3 Equation for the mechanical variables

Equations for the developed electromagnetic torque and rotating speed are given in the previous chapter:

$$t_{\mathrm{e}} = p\left(\psi_{\mathrm{d}} i_{\mathrm{q}} - \psi_{\mathrm{q}} i_{d}\right). \tag{421}$$

This electromagnetic torque covers two components of the torque:

$$t_{\mathrm{e}} = J\frac{\mathrm{d}\Omega_{\mathrm{r}}}{\mathrm{d}t} + t_{\mathrm{L}}, \tag{422}$$

where t_{L} is load torque on the shaft of the machine, including torque of the mechanical losses, J is moment of inertia of the rotating mass, and $\mathrm{d}\Omega/\mathrm{d}t$ is time varying of the mechanical angular speed.

In the motoring operation, the rotating speed is an unknown variable which is calculated from the previous equation:

$$\frac{\mathrm{d}\Omega_{\mathrm{r}}}{\mathrm{d}t} = \frac{1}{J}[t_{\mathrm{e}} - t_{\mathrm{L}}] = \frac{1}{J}\left[p\left(\psi_{\mathrm{d}} i_{\mathrm{q}} - \psi_{\mathrm{q}} i_{\mathrm{d}}\right) - t_{\mathrm{L}}\right]. \tag{423}$$

In equations of the voltage and current, there is an electrical angular speed. Therefore it is necessary to make a recalculation from the mechanical to the electrical angular speed:

$$\frac{\mathrm{d}\omega_{\mathrm{r}}}{\mathrm{d}t} = \frac{p}{J}[t_{\mathrm{e}} - t_{\mathrm{L}}] = \frac{p}{J}\left[p\left(\psi_{\mathrm{d}} i_{\mathrm{q}} - \psi_{\mathrm{q}} i_{\mathrm{d}}\right) - t_{\mathrm{L}}\right]. \tag{424}$$

This is the further equation which is solved during the investigation of the synchronous machine transients.

2.21 Transients of the synchronous machine in the dq0 system

At first the transients of the synchronous motor without damping windings are investigated, i.e., machine is without D, Q windings and on the rotor; there is only field winding f. On the stator, the three-phase winding fed by the symmetrical voltage system is distributed. It means that the zero voltage component is zero: $u_0 = 0$. Then equations for stator windings are in the form (the variables with $'$ (prime) sign mean rotor variables or parameters referred to the stator):

$$u_{\mathrm{d}} = R_{\mathrm{s}} i_{\mathrm{d}} + \frac{\mathrm{d}\psi_{\mathrm{d}}}{\mathrm{d}t} - \omega_{\mathrm{r}} \psi_{\mathrm{q}}, \tag{425}$$

$$u_{\mathrm{q}} = R_{\mathrm{s}} i_{\mathrm{q}} + \frac{\mathrm{d}\psi_{\mathrm{q}}}{\mathrm{d}t} + \omega_{\mathrm{r}} \psi_{\mathrm{d}}, \tag{426}$$

$$u'_{\mathrm{f}} = R'_{\mathrm{f}} i'_{\mathrm{f}} + \frac{\mathrm{d}\psi'_{\mathrm{f}}}{\mathrm{d}t}, \tag{427}$$

where:

$$\psi_{\mathrm{d}} = L_{\mathrm{d}} i_{\mathrm{d}} + L_{\mu\mathrm{d}} i'_{\mathrm{f}} = \left(L_{\sigma\mathrm{s}} + L_{\mu\mathrm{d}}\right) i_{\mathrm{d}} + L_{\mu\mathrm{d}} i'_{\mathrm{f}}, \tag{428}$$

$$\psi_{\mathrm{q}} = L_{\mathrm{q}} i_{\mathrm{q}} = \left(L_{\sigma\mathrm{s}} + L_{\mu\mathrm{q}}\right) i_{\mathrm{q}}, \tag{429}$$

$$\psi_0 = L_0 i_0 = L_{\sigma\mathrm{s}} i_0, \tag{430}$$

$$\psi'_{\mathrm{f}} = L_{\mu\mathrm{d}} i_{\mathrm{d}} + \left(L'_{\sigma\mathrm{f}} + L_{\mu\mathrm{d}}\right) i'_{\mathrm{f}} = L'_{\mathrm{ff}} i'_{\mathrm{f}} + L_{\mu\mathrm{d}} i_{\mathrm{d}}. \tag{431}$$

In motoring operation, the terminal voltages are known variables, and unknown parameters are currents and angular speed. To investigate the currents, it is necessary to derive expressions from Eqs. (425) and (426), and Eq. (423) is valid for speed. On the left side of Eqs (425) and (426), there are voltages transformed into the d-axis and q-axis, which were derived in the previous chapter:

$$u_{\mathrm{d}} = \sqrt{\frac{3}{2}} U_{\max} \cos\left(\omega_{\mathrm{s}} t - \vartheta_{\mathrm{r}}\right), \tag{432}$$

$$u_{\mathrm{q}} = \sqrt{\frac{3}{2}} U_{\max} \sin\left(\omega_{\mathrm{s}} t - \vartheta_{\mathrm{r}}\right), \tag{433}$$

if the cosinusoidal functions of the phase voltages a, b, c were accepted. The rotor position expressed as the angle ϑ_{r} is linked with the electrical angular speed by equation:

$$\vartheta_{\mathrm{r}} = \int \omega_{\mathrm{r}} dt + \vartheta_{\mathrm{r}0}; \tag{434}$$

eventually:

$$\frac{\mathrm{d}\vartheta_{\mathrm{r}}}{\mathrm{d}t} = \omega_{\mathrm{r}}. \tag{435}$$

In Eqs. (432) and (433), it is seen that load angle ϑ_{L}, which is defined as a difference between the position of the rotating magnetic field magnitude, represented by the expression $\omega_{\mathrm{s}} t$, and position of the rotor axis, represented by the angle ϑ_{r}, is present directly at the expression for the voltages:

$$\omega_{\mathrm{s}} t - \vartheta_{\mathrm{r}} = \vartheta_{\mathrm{L}}, \text{eventually } \frac{\mathrm{d}\vartheta_{\mathrm{L}}}{\mathrm{d}t} = \omega_{\mathrm{s}} - \omega_{\mathrm{r}}. \tag{436}$$

Because the load angle is a very important variable of the synchronous machine, its direct formulation in the voltage expressions should be employed in simulation with benefit.

For motor operation the currents, as unknown variables, are derived from the voltage equations after the linkage magnetic fluxes are introduced. In the next text, it is taken as a matter of course that rotor variables are referred to the stator and this fact is not specially marked.

$$u_d = R_s i_d + L_d \frac{di_d}{dt} + L_{df} \frac{di_f}{dt} - \omega_r L_q i_q, \tag{437}$$

$$u_q = R_s i_q + L_q \frac{di_q}{dt} + \omega_r L_d i_d + \omega_r L_{df} i_f \tag{438}$$

$$u_f = R_f i_f + L_{fd} \frac{di_d}{dt} + L_{ff} \frac{di_f}{dt}. \tag{439}$$

If from Eq. (439), $\frac{di_f}{dt}$ is expressed and introduced to Eq. (437), then after modifications it yields:

$$\frac{di_d}{dt} = \frac{L_{ff}}{L_d L_{ff} - L_{df}^2} \left(u_d - R_s i_d + \omega_r L_q i_q - \frac{L_{df}}{L_{ff}} u_f + \frac{L_{df}}{L_{ff}} R_f i_f \right). \tag{440}$$

Similarly, if from (437) expression $\frac{di_d}{dt}$ is introduced to the (439), then after some modifications it yields:

$$\frac{di_f}{dt} = \frac{L_d}{L_d L_{ff} - L_{df}^2} \left(u_f - R_f i_f - \frac{L_{fd}}{L_d} u_d + \frac{L_{fd}}{L_d} R_s i_d - \omega_r \frac{L_{fd}}{L_d} L_q i_q \right). \tag{441}$$

From Eq. (438), an expression for current in the q-axis is received:

$$\frac{di_q}{dt} = \frac{1}{L_q} \left(u_q - R_s i_q - \omega_r L_d i_d - \omega_r L_{df} i_f \right). \tag{442}$$

Consequently at transient investigation, a system of Eqs. (432), (433), (435), (437), (440), (441), (442), and (424) must be solved. The outputs are the currents of the stator windings in the form of i_d, i_q, which are fictitious currents. The real stator phase currents must be calculated by inverse transformation:

$$i_a = i_s = \frac{2}{3} \frac{1}{\sqrt{\frac{2}{3}}} i_d \cos \vartheta_r - \frac{2}{3} \frac{1}{\sqrt{\frac{2}{3}}} i_q \sin \vartheta_r = \sqrt{\frac{2}{3}} i_d \cos \vartheta_r - \sqrt{\frac{2}{3}} i_q \sin \vartheta_r. \tag{443}$$

This is current in the a-phase, and currents in the other phases are shifted about 120°. In Section 23 there is an example of a concrete synchronous motor with its nameplate and parameters, equations, and time waveforms of the investigated variables at transients and steady-state conditions.

Following the same equations, it is possible to investigate also generating operation with that difference that mechanical power is delivered, which requires negative load torque in the equations and to change sign at the current i_q. Then the electromagnetic torque is negative.

2.22 Transients of synchronous machine with permanent magnets

If synchronous machine is excited by permanent magnets (PM), this fact must be introduced into equations, which are solved during transients. See [15–17].

At first it is necessary to determine linkage magnetic flux of permanent magnets ψ_{PM}, by which an electrical voltage is induced in the stator winding. Obviously, it is measured on a real machine at no load condition in generating operation. From Eq. (73) it can be derived that in a general form, the PM linkage magnetic flux is given by equation:

$$\psi_{PM} = \frac{U_i}{\omega}.$$

In Eq. (428), there is instead the expression with a field current directly ψ_{PM}. As it is supposed that this magnetic flux is constant, its derivation is zero, and Eq. (437) is in the form:

$$u_d = R_s i_d + L_d \frac{di_d}{dt} - \omega_r L_q i_q, \tag{444}$$

$$u_q = R_s i_q + L_q \frac{di_q}{dt} + \omega_r L_d i_d + \omega_r \psi_{PMdq}. \tag{445}$$

In that equation the PM linkage magnetic flux is transformed into the dq0 system, because also the other variables are in this system. To distinguish it from the measured value, here a subscript "dq" is added. It can be determined as follows: In no load condition at the rated frequency, the currents i_d, i_q are zero; therefore also the voltage u_d is zero according to Eq. (444), and the voltage in q-axis, according to Eq. (445), is:

$$u_q = \omega_{rN} \psi_{PMdq} \tag{446}$$

and at the same time according to Eq. (418), in absolute value, is:

$$u_q = \sqrt{\frac{3}{2}} U_{max},, \tag{447}$$

because also load angle is, in the no-load condition, zero. Then:

$$\psi_{PMdq} = \sqrt{\frac{3}{2}} \frac{U_{max}}{\omega_{rN}}. \tag{448}$$

This value is introduced into Eq. (445), to calculate the currents i_d, i_q. The real currents in the phase windings are obtained by an inverse transformation according to Eq. (443). Examples of these motor simulations are in Section 23.2.

2.23 Transients of a concrete synchronous motor

2.23.1 Synchronous motor with field winding

Equations from Section 21 are used for transient simulations of a concrete synchronous motor with field winding. The nameplate and parameters of this motor are shown in **Table 5**.

Seeing that the simulation model is created in the d, q-system, linked with the rotor position, also the terminal voltages must be given in this system. It is made by Eqs. (419) and (420). The field winding voltage is a DC value and is constant during the whole simulation.

As it is known from the theory of synchronous motor, the starting up of synchronous motor is usually not possible by directly switching it across the line. If the synchronous motor has a damping winding, which is originally dedicated for damping of the oscillating process during motor operation, this winding can act as a squirrel cage and develop an asynchronous starting torque sufficient to get started. After the motor achieves the speed close to synchronous speed, it falls spontaneously into synchronism.

The damping winding is no longer active in torque development. The investigated motor has no damping winding; therefore a frequency starting up is carried out, which means continuously increasing voltage magnitude and frequency in such a way that their ratio is constant. In **Figure 40**, simulation waveforms of frequency starting up of this motor are shown, which means time dependence of variables n =

S_N = 56 kVA	R_s = 0.0694 Ω
U_N = 231 V	$L_{\sigma s}$ = 0.391 mH
I_N = 80.81 A	$L_{\mu d}$ = 10.269 mH
f_N = 50 Hz	$L_{\mu q}$ = 10.05 mH
U_f = 171.71 V	R_f' = 0.061035 Ω
I_f = 11.07 A	$L_{\sigma f}$ = 0.773 mH
n_N = 1500 $\min^{-1}$	J = 0.475 kg m^2
T_N = 285 Nm	p = 2
T_{loss} = 1 Nm	

Table 5.
Nameplate and parameters of the investigated synchronous motor with field winding.

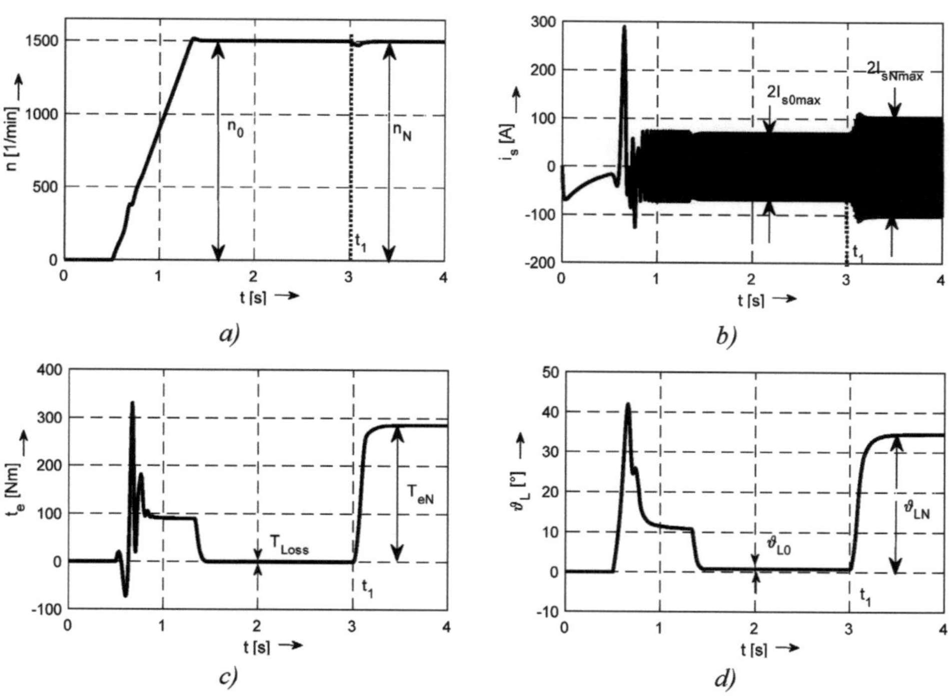

Figure 40.
Simulation time waveforms of the synchronous motor with field winding starting up: (a) speed n*, (b) current of the a-phase* $i_a = i_s$*, (c) developed electromagnetic torque* t_e*, and (d) load angle* ϑ_L*.*

f(t), i_a = f(t), t_e = f(t), and ϑ_L = f(t). Motor is at the instant t = 3 s loaded by rated torque T_N = 285 Nm.

The mechanical part of the model is linked with Eq. (423); however, it had to be spread by the damping coefficient of the mechanical movement of the rotor. This coefficient enabled implementation of the real damping, which resulted in a more stable operation of the rotor.

2.23.2 Synchronous motor with PM

Equations derived in Section 22 are employed in transient simulation of the synchronous motor with PM. The equation for the developed electromagnetic torque (221) is modified to the form:

$$t_e = p(L_d i_d + \psi_{PM}) i_q - L_q i_q i_d. \tag{449}$$

P_N = 2 kW	R_s = 3.826 Ω
U_{phN} = 230 V, star connection Y	L_d = 0.07902 H
I_N = 10 A	L_q = 0.16315 H
f_N = 36 Hz	ψ_{PM} = 0.8363 Wb
n_N = 360 min^{-1}	J = 0.02 kg m^2
T_N = 54 Nm	p = 6
T_{loss} = 0.5 Nm	

Table 6.
Nameplate and parameters of the investigated synchronous motor with PM (SMPM).

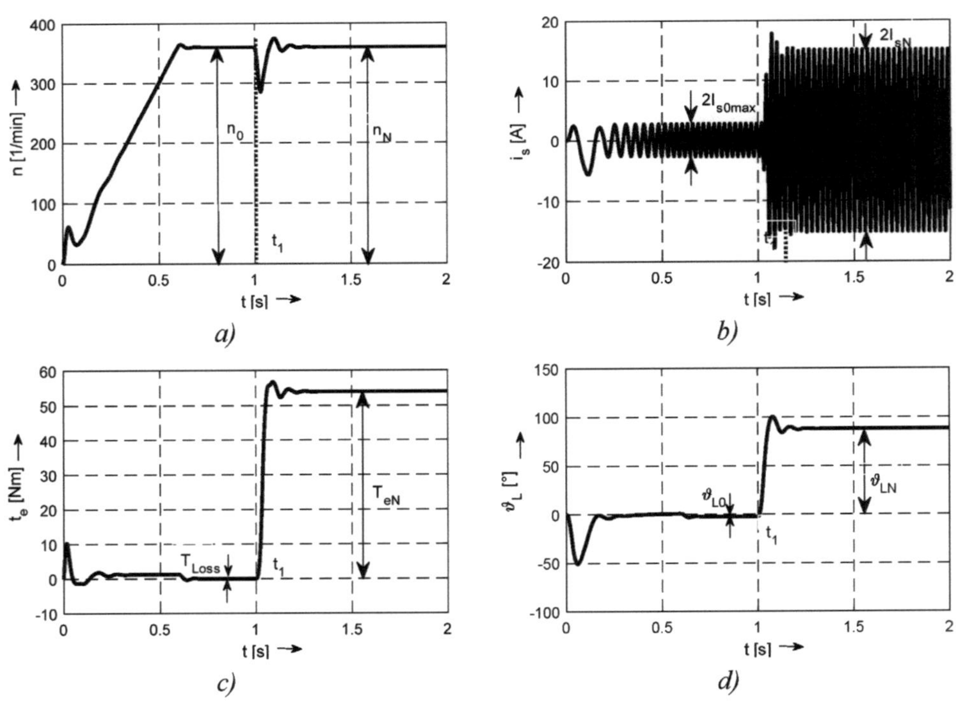

Figure 41.
Simulated time waveforms of the synchronous motor with PM: (a) speed n, *(b) current of the a-phase* i_a, *(c) electromagnetic developed torque* t_e, *and (d) load angle* ϑ_L.

Also this simulation model is created in the system d, q linked with the rotor position; therefore, it is necessary to adjust the terminal voltage to this system. The nameplate and parameters of the investigated motor are in **Table** 6.

In **Figure 41**, there are simulation waveforms of the simulated variables during the frequency starting up of the synchronous motor with PM, if stator voltage and its frequency are continuously increasing in such a way that their ratio is constant. It is true that this way of the starting up is not typical for this kind of the motors. Such motors are usually controlled by field-oriented control (FOC), but its explaining and application exceed the scope of this textbook.

Simulation waveforms show time dependence of the n = f(t), i_a = f(t), t_e = f(t), and ϑ_L = f(t). The motor is at the instant t = 3 s loaded by rated torque T_N = 54 Nm. The mechanical part is widened by damping of the mechanical movement of the rotor.

Chapter 3

Analysis of Electrical Machines Using Finite Element Method

3.1 Introduction

Design of electrical machines is a very important task from several points of their qualitative parameters. The design methods were and still are developed to obtain best results in electrical and magnetic optimal utilization of machine circuits. One of the main tasks of electrical machine design is to maximize output power and efficiency and to decrease its volume and losses. For this design, analytical and empirical terms, verified by years of experience in designing the electrical machines, are used. The best-known designs are listed in [1–3].

With the development of computer technology, however, the old-new method for analyzing the distribution of electromagnetic fields in electrical machines, for determining the parameters of the machine equivalent circuit and for calculating other properties, is also possible. It is the finite element method (FEM), which was previously derived as a mathematical method, but it has not been established until recently due to its complexity in the calculations. Its use enabled the development of computer technology. It is currently used to solve problems not only in most industries but, for example, also in medicine.

In electrical machines, FEM is used in conjunction with the numerical solution of Maxwell's equations in the analyzed electrical machine. The solution can be carried out in plane (2D) or in space (3D). In cylindrical electrical machines where the diameter of the machine is negligible in relation to the active length of the machine, it is sufficient to solve the problem in 2D, i.e., in the cross section of the machine. In some electrical machines, the deviation of the calculation is of the order of a percentage compared to measurements (reluctance machines), in some up to 10–20%, e.g., in asynchronous machines. Various types of commercial programs that use FEM can be used for the calculation, e.g., Ansys, Ansoft, Quickfield, Opera, Maxwell, etc. This chapter uses a program that is freely distributed on the web and is called the finite element method magnetostatics (FEMM), see [10].

The advantage of FEM analysis of an electrical machine is that the FEM can also solve an electrical machine that does not yet exist or is only under the design process. Therefore, it is profitable to use FEM in the design of electrical machines. First, the electrical machine is designed by years of verified design calculations and equations of the electrical machine design, if known, and then its dimensions and other parameters are fine-tuned using FEM.

The calculation of the electrical machine parameters from its dimensions and the properties of the materials used (using analytical methods and equations) take into account the distribution of the magnetic field in the machine but in a simplified way. In opposite, the FEM program uses this magnetic field distribution as the basis for its calculations, including the saturation of magnetic circuits and the magnetic flux density distribution in the air gap, taking into account the stator and rotor slots.

Therefore, the parameters of the electrical machine equivalent circuit, which are determined by FEM, should be more accurate than from the classical design calculation.

3.2 Physical basis of FEMM calculations

This section describes the physical basis of FEMM calculations [10]. For low-frequency tasks, including electrical machines that are solved in FEMM, only some Maxwell equations are sufficient. Low-frequency tasks are those tasks in which shifting currents can be neglected. The shifting currents are important at radio frequencies only. Therefore, magnetostatic analysis can be used. The solution will be based on four Maxwell equations. In some publications, Maxwell equations have various rankings, because the authors use their own labeling, e.g., [1]. Therefore, they are ranked on the basis of physical laws and their discoverers. More detailed information on the different shapes of Maxwell's equations can be found, e.g., in [12]. It can be started with the Ampère's circuital law, which is in integral form as follows:

$$\oint_{l} \boldsymbol{H} \mathrm{d}\boldsymbol{l} = I + \frac{\mathrm{d}\psi}{\mathrm{d}t} \tag{450}$$

where $\boldsymbol{H}$ is vector of magnetic field strength, $\boldsymbol{l}$ is the mean length of the field line, I is total current, and ψ is magnetic flux linkage.

Another law is the magnetic flux density law, known as Faraday's law of induction, which is defined in integral form:

$$\oint_{l} \boldsymbol{E} \mathrm{d}\boldsymbol{l} = -N\frac{\mathrm{d}\phi}{\mathrm{d}t} \tag{451}$$

where $\boldsymbol{E}$ is electric field strength, N in number of turns, and ϕ is magnetic flux.

The next equation corresponds to magnetic flux linkage law, which is defined as:

$$\oint_{S} \boldsymbol{B} \mathrm{d}\boldsymbol{S} = 0 \tag{452}$$

where $\boldsymbol{B}$ is vector of magnetic flux density of magnetic field and $\boldsymbol{S}$ is cross section, through magnetic flux passes.

The last law that belongs to Maxwell's equations is Gauss's law of electrostatics, which can be applied using finite element method only to solve electrostatic field. Its integral form is:

$$\oint_{S} \boldsymbol{D} \mathrm{d}\boldsymbol{S} = Q \tag{453}$$

where $\boldsymbol{D}$ is electric flux density and Q is amount of electric charge.

To analyze properties of electrical machines using FEM, first magnetostatic analysis is applied, in which magnetic vector potential $\boldsymbol{A}$ has an important role. The definition of magnetic vector potential value is analogous to the definition of electric field potential. The essential difference is that magnetic potential is vector, while electric potential is scalar. Enclosed conductor of length l with current I creates magnetic vector potential, which is defined as:

$$\boldsymbol{A} = \frac{\mu_0}{4\pi} \oint_{l} \frac{I \mathrm{d}\boldsymbol{l}}{r} \tag{454}$$

where r is the value of $\boldsymbol{r}$-vector, of which end point remains in place where the potential is being evaluated, while being integrated. Magnetic flux density vector $\boldsymbol{B}$ can be derived from magnetic vector potential, using derivations. We will show the derivations using the defined vector product of nabla operator and magnetic vector potential $\boldsymbol{A}$, according to [10]. Note that in Eq. (455), $\boldsymbol{r}$ is the nominator and r the denominator:

$$\nabla \times \boldsymbol{A} = \frac{\mu_0}{4\pi} \oint_l \left(\nabla \frac{I}{r} \right) \times \mathrm{d}\boldsymbol{l} = -\frac{\mu_0}{4\pi} \oint_l \left(\frac{I\boldsymbol{r}}{r^3} \right) \times \mathrm{d}\boldsymbol{l} = \frac{\mu_0}{4\pi} \oint_l \left(\frac{I\mathrm{d}\boldsymbol{l} \times \boldsymbol{r}}{r^3} \right) = \boldsymbol{B}. \tag{455}$$

This equation between magnetic vector potential $\boldsymbol{A}$ and magnetic flux density $\boldsymbol{B}$ often utilizes FEM to evaluate magnetostatic field. It will be shown in further chapters where various types of electrical machines are analyzed.

If the magnetostatic problem is solved, magnetic fields are constant in time. In this case, equations for magnetic field strength $\boldsymbol{H}$ and magnetic flux density $\boldsymbol{B}$ are defined by the following equations:

$$\nabla \times \boldsymbol{H} = \boldsymbol{J} \tag{456}$$

where $\boldsymbol{J}$ is the current density.

The relationship between B and H for each material is given by the equation:

$$\boldsymbol{B} = \mu \boldsymbol{H}. \tag{457}$$

If the material is nonlinear (e.g., saturated iron or Alnico magnets), permeability μ is the function of B:

$$\mu = \frac{B}{H(B)}, \mu = f(B). \tag{458}$$

The FEMM program analyzes the electromagnetic field; thus, it solves equations from (455) to (458) utilizing magnetic vector potential approach. Then it can be written as:

$$\nabla \times \left(\frac{1}{\mu(B)} \nabla \times \boldsymbol{A} \right) = \boldsymbol{J}. \tag{459}$$

For linear isotropic material and supposed validity of Coulomb's criterion $\nabla A = 0$, Eq. (459) is reduced to:

$$-\frac{1}{\mu} \nabla^2 \boldsymbol{A} = \boldsymbol{J}. \tag{460}$$

FEMM retains Eq. (459), allowing to solve also magnetostatic tasks with nonlinear *B-H* equation. These equations are used, when the no load condition of electrical machine is solved.

In overall three-dimensional case, vector $\boldsymbol{A}$ consisted of three components. Before we will show how to calculate forces, torques, inductances, etc., utilizing FEMM, we can say that some tasks can be solved based on the time variable field. Such approach is called harmonic task. It can be used, e.g., to evaluate asynchronous motor in no load condition. Harmonic task evaluation will be shown in Section 5.

Harmonic task can be considered, when field is time variable (e.g., harmonically changing current). In materials with non-zero permeability, eddy currents can be induced. Basic equation for electric field strength E and current density J is defined as:

$$J = \sigma E \tag{461}$$

where σ is specific conductivity.

Integral form of electromagnetic flux density law (induction law) can be rewritten as follows:

$$\nabla \times \boldsymbol{E} = -\frac{\partial B}{\partial t}. \tag{462}$$

where B is substituted with vector potential expression:

$$\nabla \times \boldsymbol{E} = -\nabla \times \boldsymbol{A}. \tag{463}$$

In the case of solving 2D problem, (463) can be rewritten as:

$$\boldsymbol{E} = -\boldsymbol{A} - \nabla \boldsymbol{V} \tag{464}$$

and by substituting to Eq. (461), it is:

$$\boldsymbol{J} = -\sigma \boldsymbol{A} - \sigma \nabla \boldsymbol{V}. \tag{465}$$

By substituting to Eq. (459), differential expression is obtained:

$$\nabla \times \left(\frac{1}{\mu(B)} \nabla \times \boldsymbol{A} \right) = -\sigma \boldsymbol{A} - \sigma \nabla \boldsymbol{V} + \boldsymbol{J}_{\mathbf{src}} \tag{466}$$

where J_{src} represents applied current sources. Component ∇V is additional voltage gradient, which is constant in 2D problem. FEMM applies this voltage gradient in some harmonic tasks to constrain current in conductive areas.

Equation (466) is considered in case, when field oscillates at one constant frequency. In such case, a steady-state equation is obtained, which is solved for amplitude and phase of vector potential A. This transformation is:

$$A = \text{Re}\left[a(\cos \omega t + j \sin \omega t)\right] = \text{Re}\left[a e^{j\omega t}\right] \tag{467}$$

in which a is a complex number. Substituting to Eq. (466) and its separation, we can get an equation, which can be used by FEMM when solving harmonic tasks:

$$\nabla \times \left(\frac{1}{\mu(B)} \nabla \times a \right) = -j\omega\sigma a - \sigma \nabla V + J_{\text{src}} \tag{468}$$

in which J_{src} represents transformed phasor of applied current sources.

For harmonic analysis, permeability μ should be constant. However, FEM takes into account nonlinear equation also in harmonic formulation, which allows solving nonlinear tasks similar to magnetostatic analysis. FEMM also includes complex and frequency-dependent permeability in time-harmonic tasks. These properties enable the program to model materials with thin sheets and also to approximately model hysteresis in ferromagnetic materials.

3.2.1 Calculation of forces and torques using FEM

Calculation of forces and torques of electrical machines is one of the most important properties of FEMM, which can be used toward this purpose.

We can use different means, but most known are these three methods of force and torque calculation:

- Maxwell stress tensor
- Coenergy method
- Lorentz force equation

3.2.1.1 Maxwell stress tensor

Using of this tensor is very simple from force and torque calculation point of view, because it requires only local distribution of magnetic field density along straight line or curve, which can be selected in the analyzed problem. In linear motion systems (e.g., linear motors), force calculation is applied. In rotating electrical machine, we can define circle in the middle of air gap, on which the electromagnetic torque of machine can be calculated. This is written in more details in Sections 5, 6 and 7. Total force can be calculated as:

$$\boldsymbol{F} = \iint \left[\frac{1}{\mu_0}\boldsymbol{B}(\boldsymbol{B}\cdot\boldsymbol{n}) - \frac{1}{2\mu_0}B^2\boldsymbol{n}\right]\mathrm{d}S \tag{469}$$

where normal component of the force is:

$$F_{\mathrm{n}} = \frac{l_{\mathrm{Fe}}}{2\mu_0}\int \left[B_{\mathrm{n}}^2 - B_{\mathrm{t}}^2\right]\mathrm{d}l \tag{470}$$

and tangential component of the force is:

$$F_{\mathrm{t}} = \frac{l_{\mathrm{Fe}}}{\mu_0}\int B_{\mathrm{n}}B_{\mathrm{t}}\mathrm{d}l \tag{471}$$

where n is normal vector, S is the area, l_{Fe} is active length of the iron in rotating electrical machines or z-axis, B_{n} is normal component of magnetic flux density, and B_{t} is tangential component of the magnetic flux density.

The torque can be calculated as follows:

$$\boldsymbol{T} = \boldsymbol{r} \times \boldsymbol{F} \tag{472}$$

Then

$$T = \frac{l_{\mathrm{Fe}}}{\mu_0}\oint_l rB_{\mathrm{n}}B_{\mathrm{t}}\mathrm{d}l \tag{473}$$

where r is radius of the circle, where the torque is calculated, and l is its length. The accuracy of this method to calculate the force or torque depends on finite elements number in the air gap of the machine, where the force or torque is calculated. The greater the number of elements, the more accurate the calculation is.

3.2.1.2 Coenergy method for force and torque calculation

Coenergy method of force or torque calculation is based on electrical to mechanical energy conversion principle in systems with variable air gap and ferromagnetic core. Among such systems are, e.g., electromagnet of contactor and electrical device based on variable reluctance principle (reluctance synchronous machine, switched reluctance machine, see Section 7 and [18–20]). Force calculation is defined by small difference (derivation) of coenergy W' with respect to small difference (derivation) of position (deflection) x in linear motion or ϑ in calculation of torque. Thus for force, it is defined as:

$$F = \frac{\mathrm{d}W'}{\mathrm{d}x} \approx \frac{\Delta W'}{\Delta x} \tag{474}$$

and the torque is defined as:

$$T = \frac{\mathrm{d}W'}{\mathrm{d}\vartheta} \approx \frac{\Delta W'}{\Delta \vartheta}. \tag{475}$$

As it was mentioned above, for calculation of one force or torque value, two FEM calculations are needed, because difference between two coenergies is required. Setup of this position or deflection step is important. When step is too big, calculated value of force or torque can be inaccurate. Setting of step must be adjusted predictably according to the problem, which is being solved. In linear motion, it can be, e.g., Δx = 1 mm, in rotating machines single mechanical degree ($\Delta\vartheta$ = 1°). Instantaneous value of torque can be calculated as:

$$T = \left.\frac{\partial W'(i,\vartheta)}{\mathrm{d}\vartheta}\right|_{\mathrm{i=const}} = -\left.\frac{\partial W'(i,\vartheta)}{\mathrm{d}\vartheta}\right|_{\psi=\mathrm{const.}} \tag{476}$$

3.2.1.3 Lorentz force equation for force and torque calculation

By using Lorentz's law to calculate force or torque, instantaneous value of torque can be obtained as function of phase induced voltage values and phase current values for three-phase system as sum of products of instantaneous values in all three phases:

$$T = \frac{1}{\Omega}\left[u_{\mathrm{iA}}(t)i_{\mathrm{A}}(t) + u_{\mathrm{iB}}(t)i_{\mathrm{B}}(t) + u_{\mathrm{iC}}(t)i_{\mathrm{C}}(t)\right] \tag{477}$$

where $u_{\mathrm{i}}(t)$ are instantaneous induced voltages of three phases, $i(t)$ are instantaneous values of the current, and Ω is mechanical angular speed.

3.2.2 Calculation of inductance by means of FEM

The method of steady-state inductance calculation is shown in this chapter. In electrical machines, it can be calculation of the self-inductance of phase, leakage inductance, armature reaction inductance, or magnetizing inductance. We can use two ways to calculate these:

- Inductance calculation for linear systems from magnetic field energy
- Inductance calculation for nonlinear systems from flux linkage

In linear systems, inductance calculation can be carried out by electromagnetic field energy evaluation, or by coenergy, because these are equal to the linear cases. It is assumed that electrical machine represents linear system, when operating in linear area of *B-H* characteristic. This means that ferromagnetic circuit is not saturated. Calculation is done by equation:

$$W = W' = \frac{1}{2}LI^2 \Rightarrow L = \frac{2W}{I^2} \tag{478}$$

where electromagnetic field energy is defined as:

$$W = \int_V \left[\int_0^B \boldsymbol{H} \cdot \mathrm{d}\boldsymbol{B} \right] \mathrm{d}V \tag{479}$$

and *V* is volume, where the electromagnetic field energy is stored.

System is considered as nonlinear, if the machine operates in nonlinear area of *B-H* characteristic. In nonlinear systems, inductance calculation can be carried out based on flux linkage evaluation, using Stokes theorem and magnetic vector potential:

$$L = \frac{\psi}{I} = \frac{\int_S \nabla \times \boldsymbol{A} \cdot \mathrm{d}\boldsymbol{S}}{I} = \frac{\oint \boldsymbol{A} \cdot \mathrm{d}\boldsymbol{l}}{I} = \frac{\oint \boldsymbol{A} \cdot J\mathrm{d}V}{I^2} \tag{480}$$

Utilization of these equations and calculation of all parameters are shown in the next chapters.

3.2.3 Procedure of FEM utilization

In general, most software for FEM analysis consists of three main parts:

- Preprocessor: preparation and evaluation of the analyzed model
- Solver (processor): assemblage of differential equation system and its solving.
- Postprocessor: analysis of results and calculation of further required parameters

All parts here are described and applied to the FEMM program.

3.2.3.1 Preprocessor

It is a mode or part of the program, where the user creates a model with finite elements. This module contains more sub-modules or other parts. At first, geometrical dimensions of model (mm or English units) are defined. It is chosen between 2D and 3D modeling. In this chapter, FEMM program is able to do only 2D analysis. This is satisfactory in most cases of electrical machines analysis, and calculation time of evaluation is significantly shorter than in 3D. If the user disposes software with 3D, it is required to regard on the manual provided by the software.

Options can be selected between planar x, y coordinates (2D) and axisymmetric system, where it can work with polar coordinates. This system is suitable to solve, e.g., cylindrical coils. In most of the electrical machines, the planar system is used, and also the z-coordinate is selected.

- *Drawing mode*: This mode is used to draw models, which will be solved. It uses specific points given by their x, y coordinates. These points are connected with lines or curves, aiming to obtain required geometrical shape. In this part of preprocessor, the model can be imported also from CAD programs in dxf format.

- *Materials definition*: In this mode, used materials of drawn models are defined by specific quantities. Each part of the model must be outlined; thus it is clearly specified which space responds to the given material. In electrical machines materials as: air, insulation, ferromagnetic materials for magnetic circuits of cores, stators and rotors (mostly defined by *B-H-* characteristic), electric current conductors are mostly made of copper or aluminum materials, ferromagnetic materials for shafts and permanent magnets, are employed. The definition of these materials is shown in each chapter, related to particular examples. FEMM program offers a library, which contains various materials, and it is possible to define these individually, according to user's requirements and needs.

- *Excitation quantities definition*: In this mode, excitation quantities are defined. These are current density, currents, or voltages, which respond to the respective parts of model (slot, winding, coil). These values must be calculated directly or obtained for a particular state of the electrical machine (e.g., no load condition).

- *Boundaries definition*: Analysis and solving of models in FEMM use two important boundary conditions: Dirichlet boundary conditions, which are defined as the constant value of magnetic vector potential $\boldsymbol{A}$ = const. This is used mostly on the surface of electrical machines, where $\boldsymbol{A} = 0$ is supposed (**Figure 42a**). Neumann boundary conditions are used for solving symmetrical machines, where we can analyze one quarter only or smaller part which replicates in the model. It is suspected for this condition that normal derivation of magnetic vector potential is zero: $\partial \boldsymbol{A}/\partial n = 0$ (**Figure 42b**).

- *Finite element mesh creation*: This comprises the FEMM method principle. As titled, the created model is split into too many parts by finite elements. This finite number of elements covers the whole model. The more complicated the shape of the model is, the higher number of elements is required. These elements are mostly triangles, yet some software can use other shapes. By first mesh generating, the program creates a certain number of elements of certain size. This number and size can be set up by the user. It is important that dependence between number and size of elements, calculation speed, and accuracy is not linear. It is not obvious to get more accurate solution by high number of small-size elements. This can be compared to the magnetization curve of ferromagnetic material, where the number of elements is on x-axis and accuracy of solution is on y-axis. This setting is to be done by the user according to the experiences and requirements of a particular model. In each of finite elements, magnetic vector potential $\boldsymbol{A}$ is calculated.

These modes are defined as preprocessor, and we can now continue to the solver part.

3.2.3.2 Solver (processor)

In this part of the program, a system of partial differential equations is created in each node of triangle, i.e. finite element, where vector magnetic potential values are

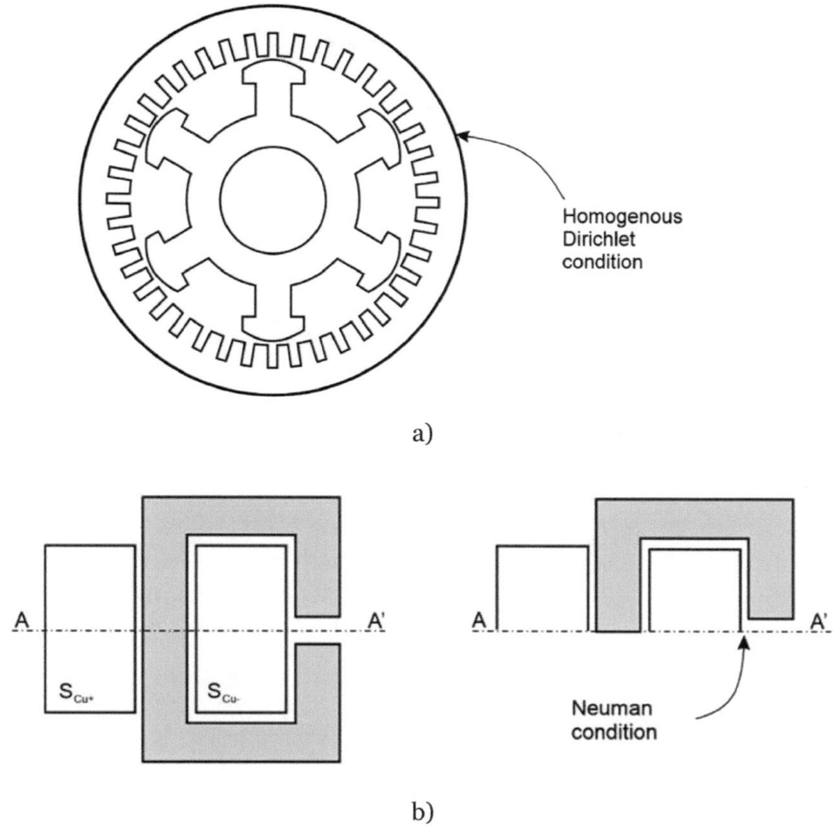

Figure 42.
Boundary conditions: (a) homogenous Dirichlet condition and (b) Neumann condition.

being solved. Based on the numerical solution of equations utilizing defined iteration methods, the program solves this system of partial differential equations. Result can converge (favorable case), where suspected solution can be obtained, or diverge, where program cannot calculate successful solution. In such case, it is necessary to revise the model and make corrections. As stated above, this part of program will take time according to the selected number and size of finite elements.

3.2.3.3 Postprocessor

In this last part, results of FEMM can be analyzed. The first result is the distribution of magnetic field in the model by means of magnetic field flux line mapping (equipotential lines). This mode allows to calculate other quantities on a particular point, defined line, or curve (by means of space integral calculation). Thus, the following can be obtained: magnetic flux density (normal and tangential component), magnetic field strength, potential, torques, energies, inductances, losses, etc.

According to this description, FEMM program will be used to analyze and calculate parameters in individual electrical devices and electrical machines. We will start with Section 3 in this chapter, where we calculate the force and inductance of a contactor electromagnet. In Section 4, we will calculate equivalent circuit parameters of single-phase transformer (mostly inductances). In Section 5, we will show calculation of parameters and torques of asynchronous machines. In Section 6, synchronous machines are analyzed, and finally in Section 7, switched reluctance machines are analyzed.

3.3 Analysis of electromagnet parameters

It is shown here how the parameters of electromagnet, or more precisely "of ferromagnetic circuit with variable air gap, fed by DC voltage," can be investigated by means of the FEMM. The purpose of this is to inform the reader how practical calculations and analyses by means of FEMM are carried out.

Such circuits are employed in electrical engineering praxis, as it is electromagnet of the relay, or rotating electrical machine with variable reluctance (Section 7). Calculation of the phase inductance and developed electrodynamic force for one general condition (moving armature is in certain position) is shown. There are two important positions of the electromagnet: switched-on position with the minimal air gap and switched-out position with maximal air gap. Examples of such circuits are in **Figure 43b,c**.

Expressions from Sections 2.1 and 2.2 are employed at the inductance and force calculations. Section 2.3 is the base for calculation procedure. At the first step (preprocessor), the base settings are: Planar Problem Type and Length Units (mm). Frequency is zero, because a magnetostatic problem is investigated, and there is given DC voltage. Set z-coordinate, i.e., active length of iron, marked in the program as Depth (**Figure 44**). After these settings, the model is drawn in such a way

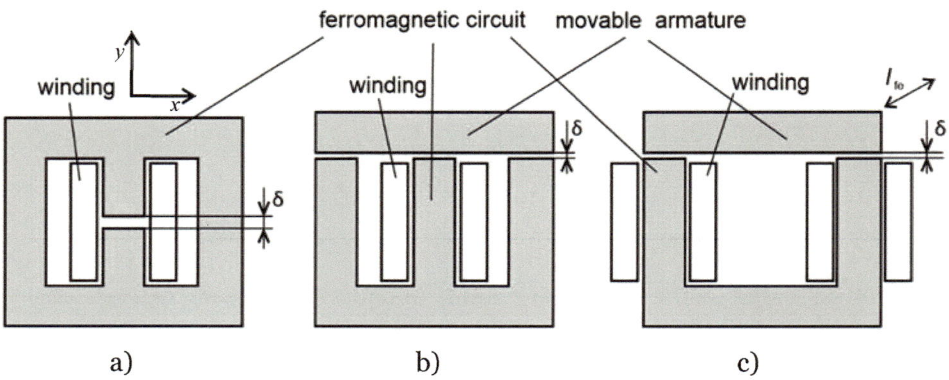

Figure 43.
Ferromagnetic circuits with an air gap: (a) constant air gap and (b, c) variable air gap.

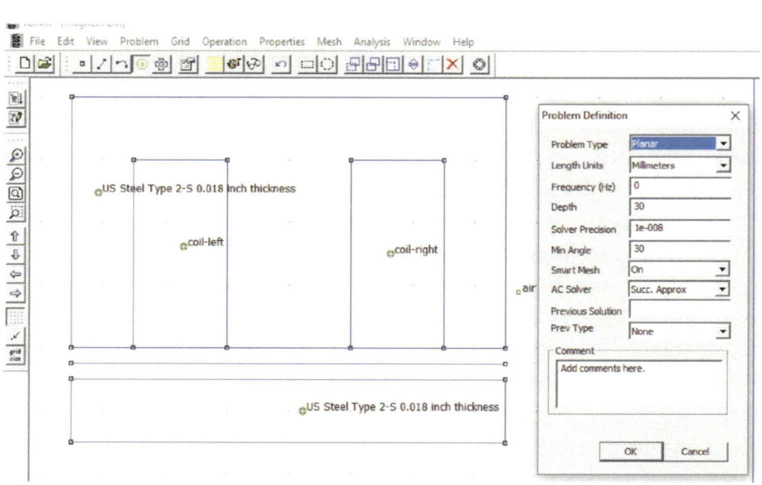

Figure 44.
Drawn mode with demonstration of the basic settings.

that the points are put in the *node mode* and by means of tabulator the coordinates x, y of the points are put in. After the points are drawn, the program can be changed to the *segment mode* and join the points to get the required shape of the ferromagnetic circuit. Into the center of the air gap, an auxiliary line is drawn, on which a value of the developed force is investigated (see **Figure 44**).

After the *drawing mode* is ready, the materials are entered. Here three materials are present: air, copper (conductors of the coil), and ferromagnetic material with the known *B-H* curve. These materials can be entered by the *Materials Library*, from which the basic material constants can be copied by the employed materials here (**Figure 45**).

Materials as air or conductor material (copper, aluminum) have relative permeability in x- and y-axis equal 1, $\mu_x = 1$, $\mu_y = 1$, and then set up electrical conductivity of a conductor (e.g., the electrical conductivity for copper is 58 MS/m). If the winding is excited by a field current, then set up the excited variables, e.g., current density of the winding. Choose, e.g., the field current $I = 2.25$ A. The number of coil turns is $N = 1000$. Cross-section area of the coil is (50 × 9) mm. The setting up is made as follows: According to the current value I and cross-section area of the conductor, calculate the current density:

$$J = \frac{NI}{S} = \frac{1000 \cdot 2.25}{0.05 \cdot 0.009} = 5\ \text{MA/m}^2. \tag{481}$$

In the FEMM program, the current density is given in MA/m^2. There is also the other way where only the number of turns and current value is entered. These settings can be made by *Properties/Circuits*. In the case of ferromagnetic materials, it is possible to take from the library a material present there with its *B-H* curve (**Figure 46**). In the case here, a *B-H* curve from the library with indication US Steel Type 2-S 0.018 inch is employed.

If all materials are defined, allocate them to drawn objects in such a way that the program is switched over to the mode *Block*, and mark gradually the objects. In this case, these are the next parts: air, right part of the coil, left side of the coil, fixed part of the ferromagnetic circuit, and moving part of the ferromagnetic circuit (armature). Materials are allocated to these parts by means of the key *space* and from the open window take a correspondent material (see **Figure 45**).

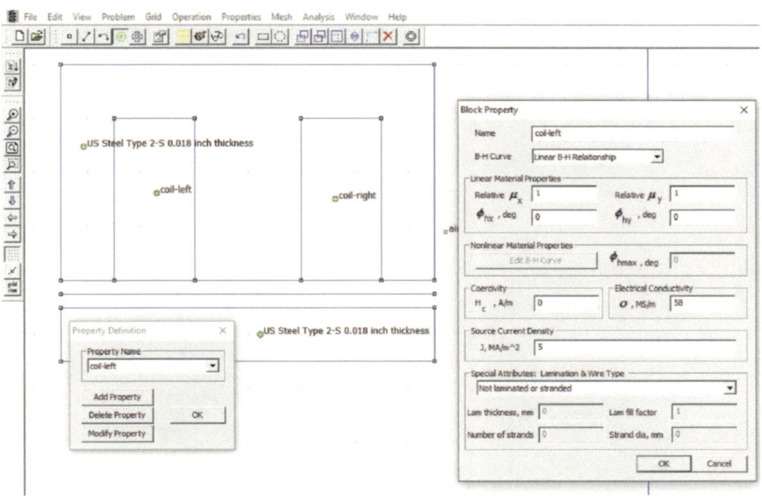

Figure 45.
Parameter setting of the materials.

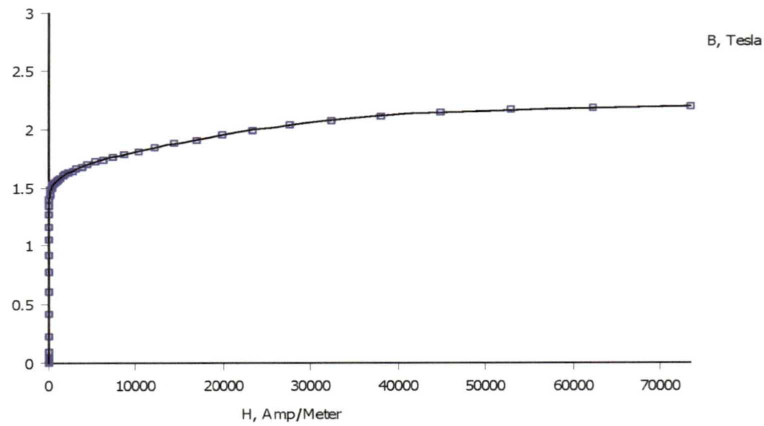

Figure 46.
The B-H curve of a ferromagnetic material taken from the Materials Library: US Steel Type 2-S 0.018 inch.

The last but one step in preprocessor is the definition of the boundary conditions. In this task, the homogenous Dirichlet condition can be chosen, if the investigated electromagnet is limited by the boundaries, where zero magnetic vector potential can be assumed. The setting up is made in *Properties/Boundary/ Prescribed A*, whereby all values are zero (**Figure 47**).

The last step in preprocessor is generation of the finite element mesh. By means of the command *Mesh/Create Mesh*, the program generates a random mesh with a relatively low number of the finite elements, which can have an important influence on the accuracy of investigation. This mesh can be refined if the correspondent material is marked and set up *Mesh size* (**Figure 48**). If, e.g., **Figure 1** is chosen, it means that the size of a finite element triangle is 1 mm, which is less than that created by an automatic program.

This way, how to set up a size of the finite elements in all materials, eventually all regions of the analyzed task, can be employed.

In this way, the preprocessor was finished, and a solver, eventually processor for construction and solving a system of partial differential equations, can be started. This can take certain time depending on the number of finite elements and complexity of the investigated circuit.

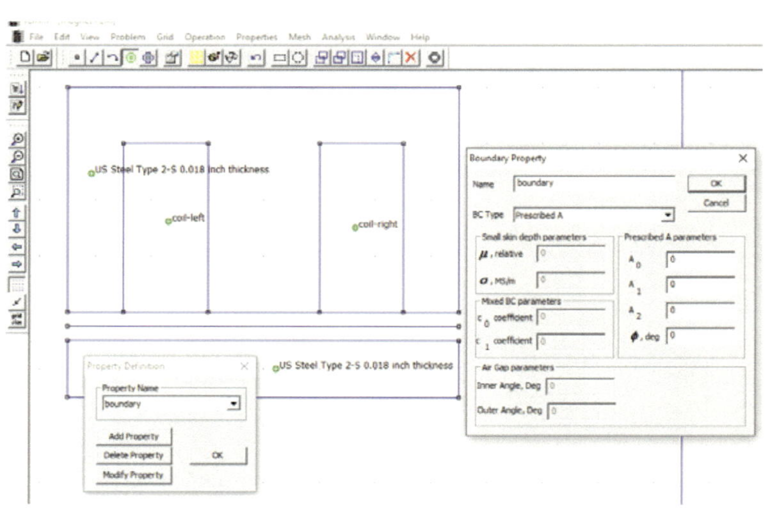

Figure 47.
Definition of the boundary condition.

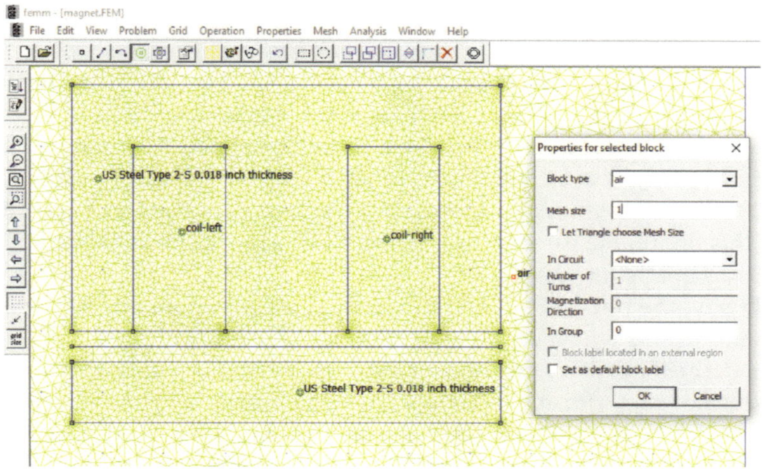

Figure 48.
Definition of the finite element size in materials.

After the solving is finished, the calculated values can be analyzed. The program shows distribution of the force lines and offer three ways how to present the outputs of the calculation:

a. In the exactly defined point (**Figure 49**) (e.g., in the point with coordinates x, y, black point). In this case a magnetic vector potential A; magnetic flux density (absolute value $|B|$, the values in direction of the x-axis, B_x, and y-axis, B_y); magnetic field intensity (absolute value $|H|$, values in x-axis, H_x, and y-axis, H_y); relative permeability in x-axis, μ_x, and in y-axis, μ_y; circuit energy E; and current density J are obtained. Because the point in which the calculation was made is in ferromagnetic material, the current density is zero: $J = 0$.

b. By means of integration along the defined line or curve. In this case, it is force calculation in the air gap.

c. By means of the marked area integration. In this case, it is inductance calculation of the magnetic circuit.

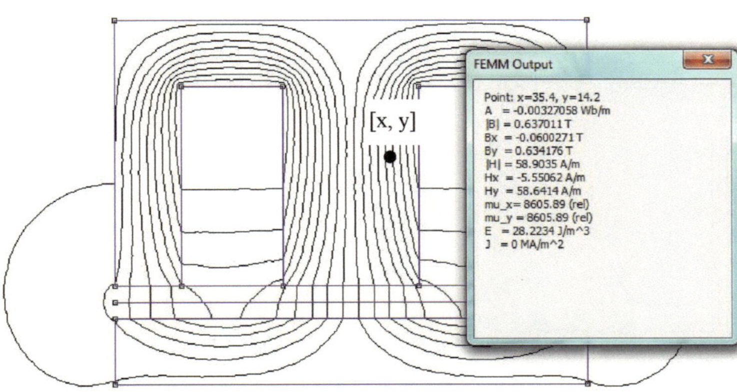

Figure 49.
Presentation of the results in the defined point x, y.

3.3.1 Electromagnet force calculation

In this case the calculation is made by means of the Maxwell stress tensor based on the equations in Section 2.1. An auxiliary line is marked, which is in the middle of the air gap between the fixed part of the electromagnet and its armature. The force is calculated by means of integration as follows: *Integrate/Line Integrals/Force form Stress Tensor*. The outputs are presented in both directions x and y (**Figure 50**). As it is seen in this figure, the force acts mainly in the y-axis direction, while in the x-axis direction, it is nearly neglected.

By this way, the force value can be investigated for various size of the air gaps and various currents which correspond to various feeding voltages.

On the defined auxiliary line, it is possible to obtain also other variables, which can help during the performance analysis, e.g., a normal component of magnetic flux density, eventually magnetic flux density in y-axis (**Figure 51**), whereby the value 0 corresponds to the length axis to the origin of the auxiliary line in the air gap and the value 70 corresponds to the end of the auxiliary line. This possibility is used in the next chapters. In the figure, there is this component that is always marked on the right side in the form B.n (marking), Tesla (here are units—see **Figure 51**), because in this form the figure is generated from the program.

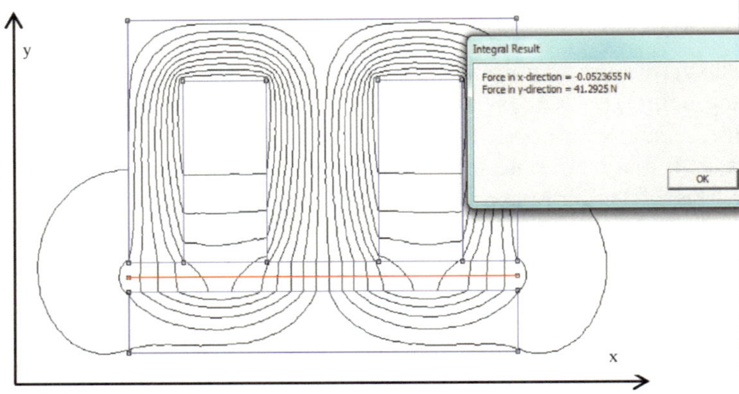

Figure 50.
Output presentation: Values of the force obtained by integration in the x-direction and y-direction on the auxiliary line in the middle of the air gap.

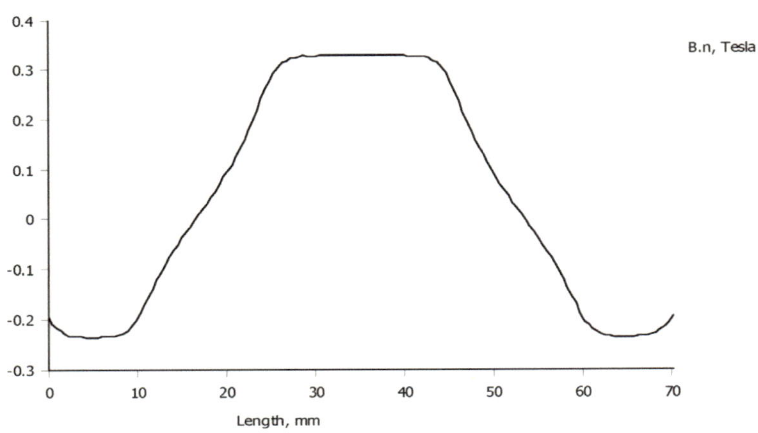

Figure 51.
Illustration of the normal component of the magnetic flux density on the defined auxiliary line.

3.3.2 Electromagnet inductance calculation

The calculation is made by two ways, as it was described in Section 2.2: by means of coenergy (linear case is supposed) and by means of linkage magnetic flux.

1. Calculation by means of coenergy

   In the program, all areas of the materials are colored by green (**Figure 52**). By means of *Integrate/Block Integrals*, calculate the *magnetic field coenergy*. Here the other calculation is also made, to be able to compare the inductances obtained by both ways. In **Figure 52** it is seen that the calculated value of coenergy is $W = 0.416682$ J.

   This value is introduced into Eq. (478) and it is received:

$$W = W' = \frac{1}{2} L I^2 \Rightarrow L = \frac{2W}{I^2} = \frac{2 \cdot 0.416682}{2.25^2} = 0.164 \text{ H}$$

   As it was written before, the electromagnet coil was fed by the current $I = 2.25$ A.

2. For the calculation by means of the linkage magnetic flux, Eq. (480) is used. Here only the blocks that correspond to the coil (**Figure 53**) are highlighted,

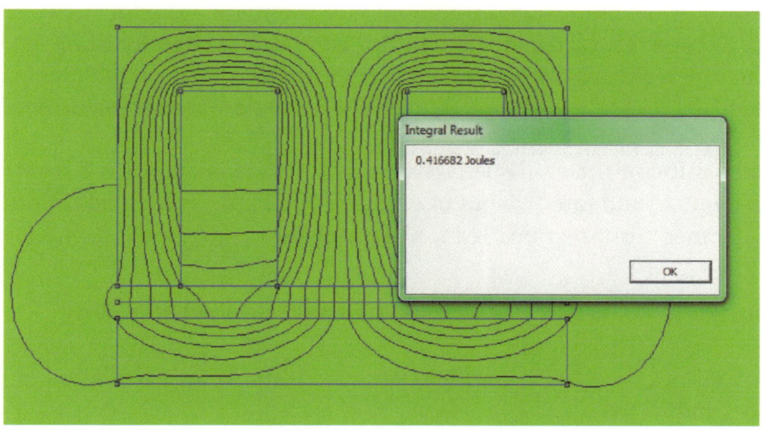

Figure 52.
Inductance calculation by means of coenergy.

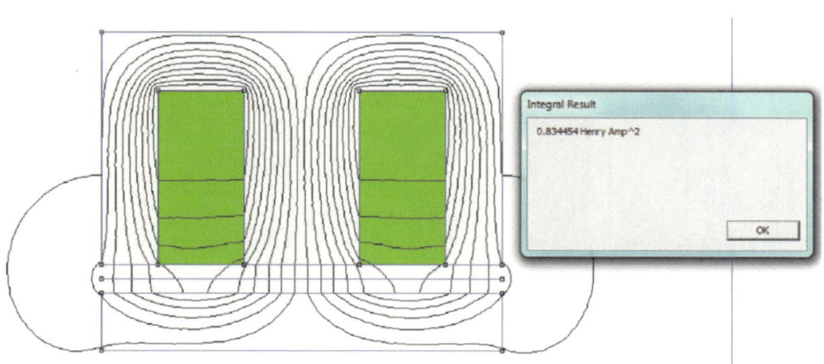

Figure 53.
Inductance calculation by means of linkage magnetic flux.

and by means of the surface integral, the value of the A.J is calculated: It is 0.8334454 HA^2 (Henry Amperes^2). Then the inductance can be calculated as:

$$L = \frac{\psi}{I} = \frac{\int_S \nabla \times \boldsymbol{A} \cdot \mathrm{d}\boldsymbol{S}}{I} = \frac{\oint \boldsymbol{A} \cdot \mathrm{d}\boldsymbol{l}}{I} = \frac{\oint \boldsymbol{A} \cdot J \mathrm{d}V}{I^2} = \frac{0.834454}{2.25^2} = 0.164 \text{ H}$$

It is seen that the results obtained by both ways of calculation are identical. If the ferromagnetic circuit was saturated, the results would be different.

Other calculations can be made for any armature position and any current. It depends on the reader needs.

3.4 Analysis of the single-phase transformer parameters

The FEM is used for the analysis of the single-phase transformer parameters. As it is known, the single-phase transformer can be designed into two main configurations: core-type construction (**Figure 54a**) and shell-type construction (**Figure 54b**). In the transformers of small powers and low voltage, the primary winding can be wound on the core and on it the secondary winding, as it is seen in **Figure 54**. Transformers for higher voltage have usually the secondary winding wound closer to the core and on it the primary winding.

The no load test can be simulated by means of FEM. The parameters of the square branch of the equivalent circuit, mainly magnetizing inductance, can be calculated. Second, also the short circuit test can be simulated by FEM. The parameters of the direct axis of the equivalent circuit, mainly leakage inductances, can be calculated. The usual equivalent circuit of the single-phase transformer is in **Figure 55**.

The calculation of the equivalent circuit parameters is made for a real transformer, nameplate and rated values of which are in **Table 7**. An illustration figure of ¼ of transformer cross-section area is in **Figure 56**.

3.4.1 Simulation of the single-phase transformer no load condition

The purpose of the simulation in no load condition is the calculation of the magnetizing inductance L_μ.

The procedure is the same as in the case of electromagnet analysis. The calculation is started with a preprocessor, where the magnetostatic analysis and the planar

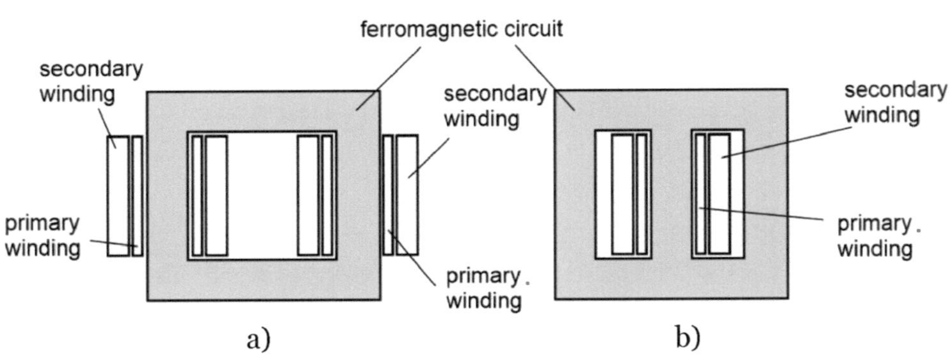

Figure 54.
Winding and core arrangements of the single-phase transformers: (a) core-type construction and (b) shell-type construction.

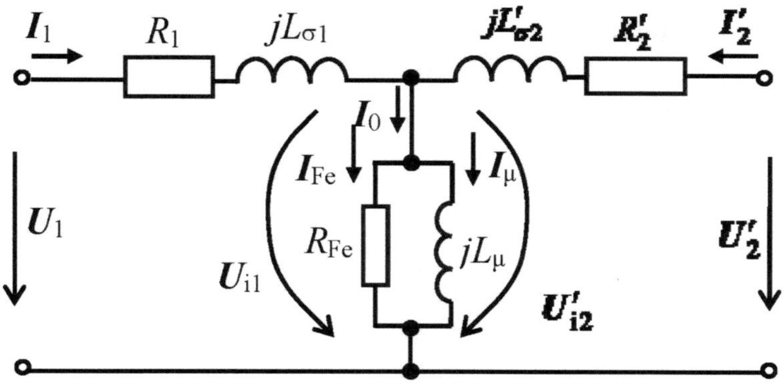

Figure 55.
Single-phase transformer equivalent circuit.

Rated voltage of the primary side U_{1N}	230 V
Rated voltage of the secondary side U_{2N}	24 V
Rated power S_N	630 VA
Rated frequency f	50 Hz
Rated current of the primary side I_{1N}	2.75 A
Rated current of the secondary side I_{2N}	26.3 A
Number of turns of the primary side N_1	354
Number of turns of the secondary side N_2	39
Shell-type construction	—
Total thickness of the core $l_{Fetotal}$	49.6 mm
Transformer sheets EI 150 N	—
No load current I_0 obtained from no load measurement	0.274 A
Magnetizing inductance L_μ obtained from no load measurement	2.82 H
Total leakage inductance L_σ obtained from short circuit measurement	5.2 mH

Table 7.
Nameplate and rated values of the investigated transformer.

type of the problem is set up. Frequency is zero, because only one time instant is solved. The z-coordinate (*depth*) is an active thickness of ferromagnetic core. The total thickness of the ferromagnetic core is obtained from the measurement $l_{Fetotal}$ = 49.6 mm. This must be reduced by the value of sheet insulation thickness and air layers between them. This reduction is respected by correction factors $k_{Fe} = l_{Fe}/l_{Fetotal} = 0.866$ - > l_{Fe} = 43 mm, and this value was used in the calculation.

Based on the transformer dimensions, a cross-section area is drawn and the materials are allocated to the blocks as follows: air around the coils; ferromagnetic material of the core; and primary and secondary windings. These are divided into two parts, right and left, and their indication is as follows: primary winding right side (pwr), primary winding left side (pwl), secondary winding right side (swr), and secondary winding left side (swl). The geometry of the cross-section area with the marked parts and setting up of the parameters are in **Figure 59**. Magnetic energy W_m is absorbed mainly into ferromagnetic core; therefore it is suitable to increase density of the mesh nodes in the core. In the other parts, the magnetic energy is neglected.

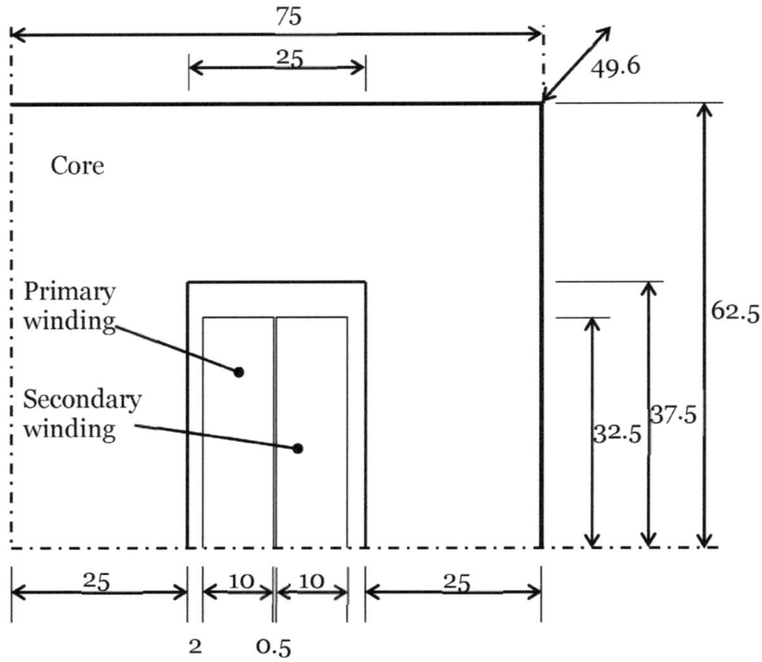

Figure 56.
An illustration figure of ¼ of transformer cross-section area.

In the no load condition, the secondary winding is open-circuited, and rated voltage at rated frequency is applied to the primary terminals. Under this condition, the primary current, the so-called no load current I_0 (**Figure 55**), flows in the primary winding. It is not necessary to draw the individual turns but the whole coil side is replaced by one block in the FEMM program. It is seen in **Figure 59** that the block (pwl) corresponds to all conductors of the left side of the primary winding. A calculation based on the magnetostatic analysis is very popular, and many authors recommend this kind of simulation [11].

A constant value of the no load current, which is the magnitude of the sinusoidal waveform, is entered in this analysis. The secondary winding is opened, in which no current flows ; therefore, a zero value of the current density is allocated to the blocks corresponding to the secondary winding. The value of the current density in the primary winding can be obtained from the analytical calculation during the transformer design or by measurements on a real transformer, which is this case.

From the no load test, the no load current at rated voltage U_N is I_0 = 0.274 A, which is an effective (rms) value. Its magnitude at the sinusoidal waveform is $I_{0\,max} = \sqrt{2}I_0 = 0.387A$, but it is true that the waveform of the no load current at rated voltage is not sinusoidal, which is caused by the iron saturation [1, 13]. For illustration in **Figure 57**, there is no load current waveform of the analyzed transformer.

The current density J_p of the magnitude of the no load current in the primary winding is calculated based on the primary winding number of turns:

$$J_p = \frac{N_1 I_{0\,max}}{S_p} = \frac{354 \cdot 0.387}{0.00065} = 0.210766\ \frac{MA}{m^2}$$

where S_p is a surface of one part of primary winding, right or left side (pwr or pwl), and can be calculated by means of the geometrical dimensions (**Figure 59**).

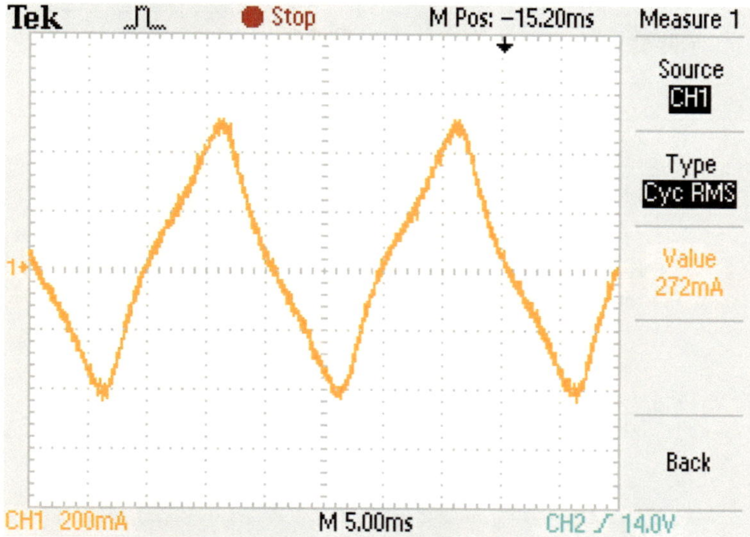

Figure 57.
No load current waveform of the analyzed transformer.

Then the current density is introduced to the block which corresponds to the right side of primary winding (pwr) $J_{pwr} = +0.210766\ MA/m^2$ and to the block corresponding to the left side of the primary winding (pwl) $J_{pwl} = -0.210766\ MA/m^2$. In the section where materials are defined, *B-H* curve is introduced, which was obtained by a producer, by the values of magnetic flux density, and by the magnetic field intensity of the employed sheets. The *B-H* curve is shown in **Figure 58**.

After a definition of all geometrical dimensions, materials, and current densities, it is necessary to define boundary conditions (**Figure 59**). Here Dirichlet boundary conditions can be applied, where constant value of the magnetic vector potential $\boldsymbol{A} = 0$ is defined. It is defined on the circumference of the transformer, and the dialog window of the FEMM program is given as *Boundary Property*, as seen in **Figure 60**.

In the last step before starting the calculation, it is necessary to create a mesh of the finite elements. As was mentioned before, in the ferromagnetic core, it is necessary to create a denser mesh, because there is concentrated dominant part of electromagnetic energy. The mesh is created by means of the command

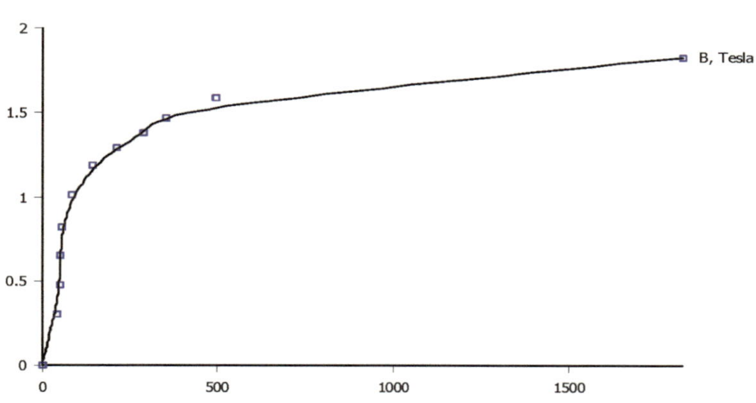

Figure 58.
B-H *curve of the sheets employed in the investigated transformer, given by its producer.*

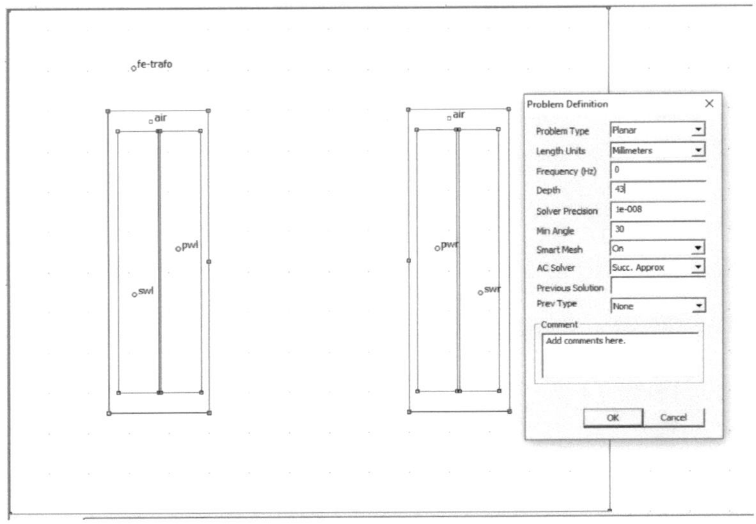

Figure 59.
Geometry of the investigated transformer with marked blocks and settings of the parameters.

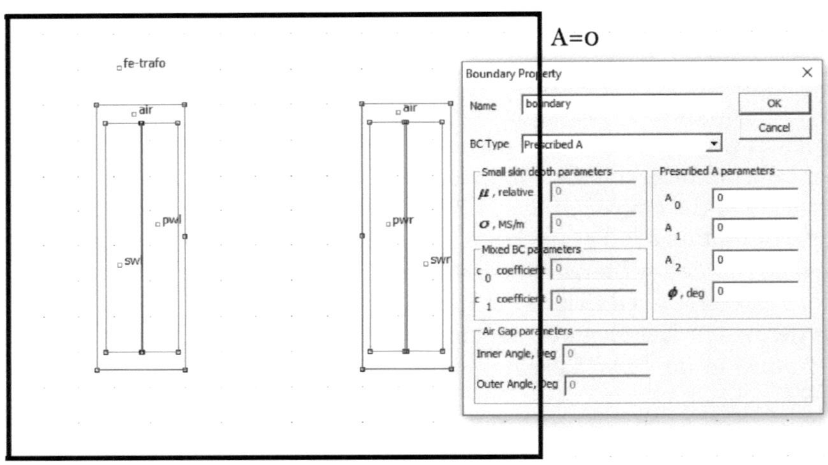

Figure 60.
Definition of the transformer boundary conditions.

Mesh – Create mesh. After a modification, there were 22,158 finite elements created (see **Figure 61**).

Now the *Processor* can be started, and calculation of the transformer in the no load condition is launched. After the calculation, a distribution of magnetic flux lines in the cross-section area of transformer is seen in *postprocessor*. Also distribution of magnetic flux density by means of command *View density plot* can be displayed (see **Figure 62**).

A magnetizing inductance calculation is done by means of linkage magnetic flux, according to Eq. (480). It must be done this way, because in the no load condition at rated voltage, the *B-H* curve is in nonlinear region, which means that energy and coenergy is not the same [4], because usually coenergy is higher, e.g., in this case the electromagnetic energy is 0.163 J and coenergy 0.263 J. It corresponds also to the non-sinusoidal waveform of the no load current, as it is seen in **Figure 57**.

Therefore, it is better to employ Eq. (480), in which linkage magnetic flux appears. In the program FEMM value A.J, is obtained in such a way that only the

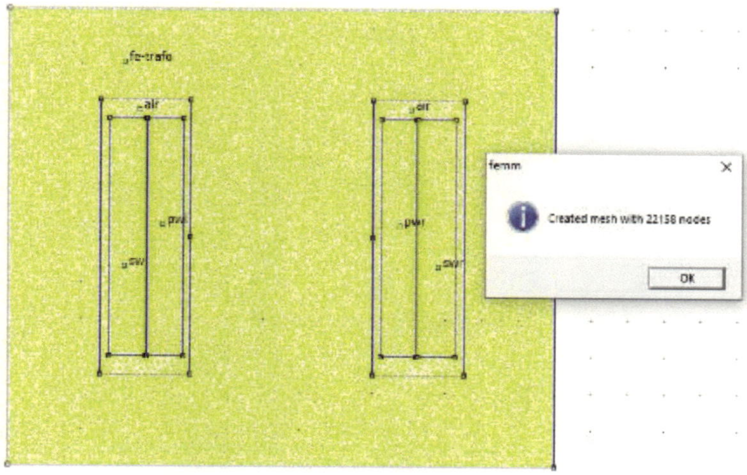

Figure 61.
Created mesh of the finite elements.

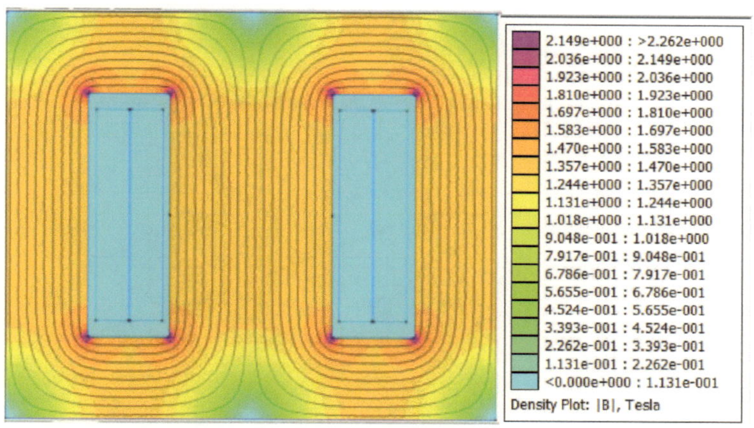

Figure 62.
Magnetic flux lines and magnetic flux density distribution in the cross-section area of the transformer.

blocks corresponding to the coils of primary winding are marked (**Figure 63**) and by means of surface integral calculate the A·J = 0.425975 H/A^2:

$$L_{\mu\text{FEMM}} = \frac{\psi}{I_{0\,max}} = \frac{\int_S \nabla \times \boldsymbol{A} \cdot d\boldsymbol{S}}{I_{0\,max}} = \frac{\oint \boldsymbol{A} \cdot d\boldsymbol{l}}{I_{0\,max}} = \frac{\oint \boldsymbol{A} \cdot \boldsymbol{J} dV}{I_{0\,max}^2} = \frac{0.425975}{0.387^2} = 2.84 \text{ H}.$$

If this value is compared with the value obtained by no load measurement (L_μ = 2.82 H, see **Table 7**), it is seen that the difference is less than 1%.

As it is known, during the measurements, it is possible to make measurement also at other voltages, not only rated, which results in various no load currents vs. voltage and also magnetizing inductances vs. no load currents. Such characteristics (L_μ vs. I_0) can be employed in transient investigation during the no load transformer switch onto the grid. For comparison in **Figure 64**, there are waveforms of such characteristics obtained by measurements and simulations. In the region of the rated voltage and corresponding no load currents I_{0N}, the coincidence is almost perfect, but at the lower currents, the discrepancy is higher because of lower saturation and perhaps lower precision of the *B-H* curve.

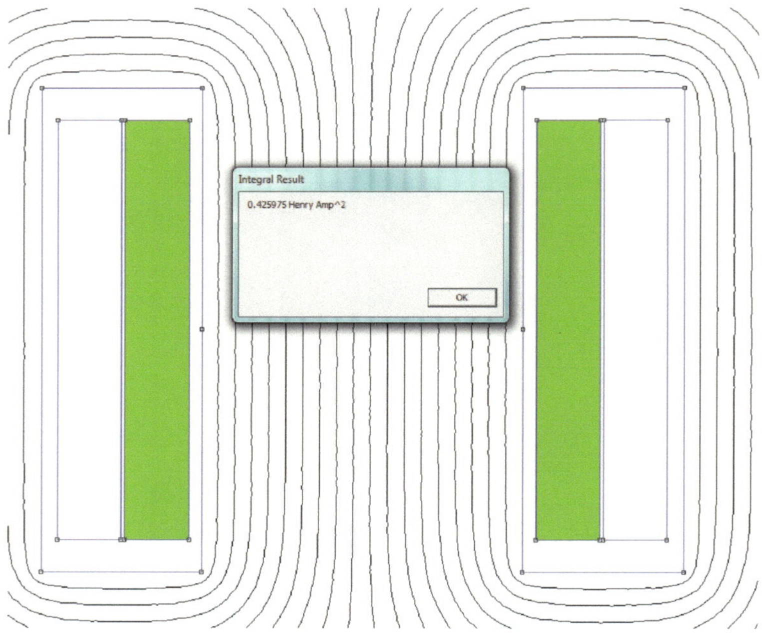

Figure 63.
Distribution of the magnetic flux lines and calculation of the surface integral A.J (integral result) of the investigated transformer.

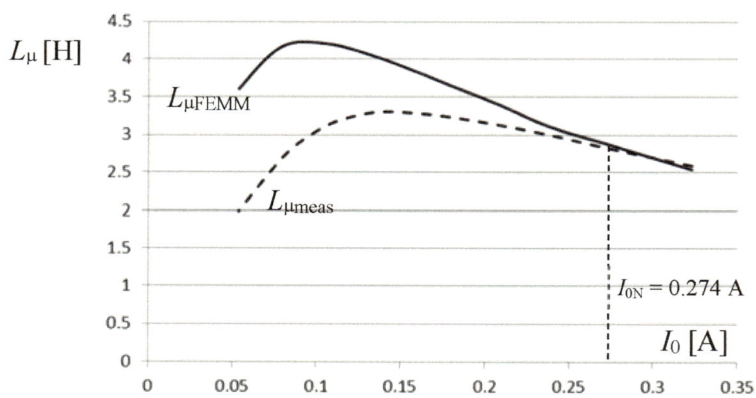

Figure 64.
Comparison of the magnetizing inductances obtained by measurement and FEMM simulation.

3.4.2 Simulation of single-phase transformer short circuit condition

According to the theory of electrical machines, the short circuit test is made at transformer, if the secondary terminals are short circuited and a fraction of the rated voltage sufficient to produce rated currents, at rated frequency is applied to the primary terminals. At the simulation of this condition, both windings are fed by their rated currents; it means that magnetomotive forces are equal and current density corresponds to the values of currents and the next equation, at which the magnetizing current is neglected $I_\mu \approx 0$ A, is valid, $N_1I_1 = -N_2I_2$, which results in the fact that no magnetizing flux is created in the core. Then only leakage flux occurs, which is closed through the leakage paths, which means by air, insulation, and nonmagnetic materials [1]. The 2D analysis is done in a different way in

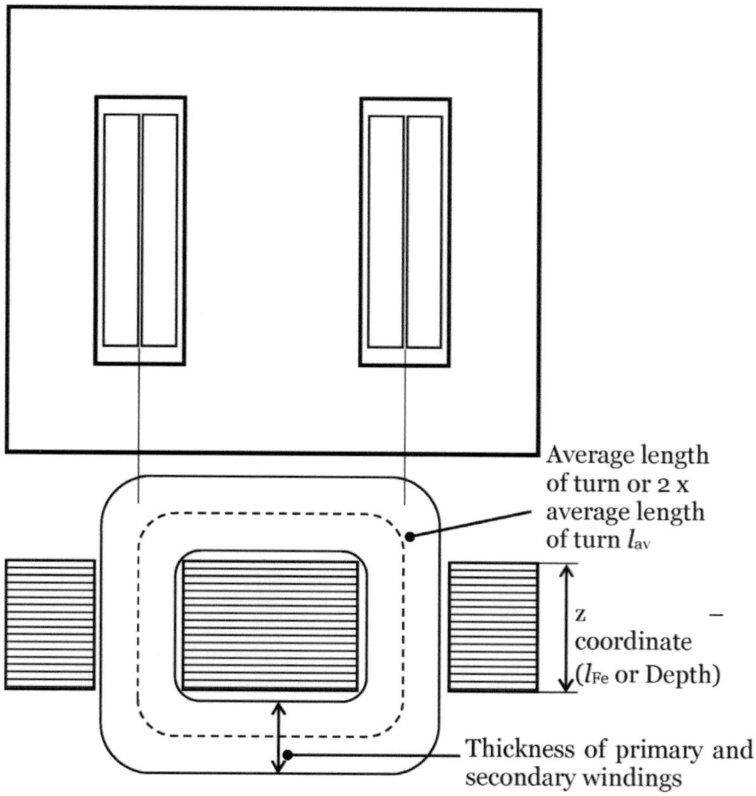

Figure 65.
Illustration figure of the transformer cross-section area to define the average length of the turn.

comparison with no load condition, when the whole magnetic flux was in the ferromagnetic core.

A procedure described in [11] is applied in the analysis of the short circuit condition. The value of the leakage inductance is calculated from the magnetic field energy, because in short circuit condition no saturation of ferromagnetic circuit occurs. The value of the energy is calculated as follows:

$$W = l_{\mathrm{av}} \int_S \frac{1}{2} \mu H^2 \mathrm{d}S$$

where S is a surface of the whole transformer cross-section area and l_{av} is an average length of the conductor or a half of the average length of the turn. It can be calculated as an average length between both windings of the primary and secondary coils (**Figure 65**). Nevertheless, based on experience, this calculated value should be increased about 5 till 10%, because during the manufacturing, the coils are not wound exactly and this dimension is very important for leakage inductance calculation. An increase of about 7.5% is used here. Based on **Figures 56** and **65**, this value can be calculated as $l_{av} = 1.075 \cdot (50 + 2 \cdot 2 + 2 \cdot 10 + 2 \cdot 0.25 + 49.6 + 2 \cdot 2 + 2 \cdot 10 + 2 \cdot 0.25) = 159.7$ mm, and the measured value is $l_{av} = 161$ mm, which is employed during further calculation. The calculation of the leakage inductance can be made by two ways:

1. The first one is setting of the z-coordinate in *Problem (Depth)* on the value l_{av}. The calculation is then made as follows:

$$L_{\sigma 1} + L_{\sigma 2}' = L_{\sigma} = \frac{2W}{{I_{1N}}^2}$$

or similarly as in the no load condition:

$$L_{\sigma 1} + L_{\sigma 2}' = L_{\sigma} = \frac{\psi}{I_{1N}^2} = \frac{\oint \boldsymbol{A} \cdot J\mathrm{d}V}{I_{1N}^2}$$

2. The second approach is such, that z-coordinate is set in *Problem (Depth)* on the value 1 mm and the values obtained in the *postprocessor* must be multiplied by the average length of the conductor or by one half of the average length of the coil turn l_{av}. Then the calculation is as follows:

$$L_{\sigma 1} + L_{\sigma 2}' = L_{\sigma} = l_{\mathrm{av}} \frac{2W}{{I_{1N}}^2}$$

or

$$L_{\sigma 1} + L_{\sigma 2}' = L_{\sigma} = l_{\mathrm{av}} \frac{\psi}{I_{1N}^2} = l_{\mathrm{av}} \frac{\oint \boldsymbol{A} \cdot J\mathrm{d}V}{I_{1N}^2}.$$

The purpose of this simulation is to calculate the total value of the leakage inductance $L_{\sigma 1} + L_{\sigma 2}' = L_{\sigma}$. In transformers that have only small number of the turns on the secondary side, e.g., there is only one layer, it is needed to draw individual turns and, in each turn, to define current or current density. If that few turns are replaced by only one turn to simplify it, a significant error could appear, because of leakage flux is flowing around the individual turns. In investigated transformer here, the secondary side is created by some layers; therefore the solution is made by one block of the turns.

A procedure of the calculation is as follows: The current densities J, corresponding to the windings, are calculated. It must be valid $N_1 I_1 = -N_2 I_2$. Then the current density is equal in both windings, but with opposite signs. The rated current in the primary winding is $I_{1\mathrm{N}} = 2.75$ A, and current density for the windings is as follows:

Primary winding, left side:

$$J_{\mathrm{pwl}} = \frac{N_1 I_{1N}}{S_{\mathrm{p}}} = \frac{354 \cdot 2.75}{0.00065} = 1.497693\ \mathrm{MA/m^2}$$

Primary winding, right side:

$$J_{\mathrm{pwr}} = -\frac{N_1 I_{1N}}{S_{\mathrm{p}}} = -\frac{354 \cdot 2.75}{0.00065} = -1.497693\ \mathrm{MA/m^2}$$

In coincidence with the equation $N_1 I_1 = -N_2 I_2$, the secondary winding, left side, is:

$$J_{\mathrm{swl}} = -\frac{N_1 I_{1N}}{S_{\mathrm{p}}} = -\frac{354 \cdot 2.75}{0.00065} = -1.49769\ \mathrm{MA/m^2}$$

and secondary winding, right side, is:

$$J_{\mathrm{swr}} = \frac{N_1 I_{1N}}{S_{\mathrm{p}}} = \frac{354 \cdot 2.75}{0.00065} = 1.49769\ \mathrm{MA/m^2}$$

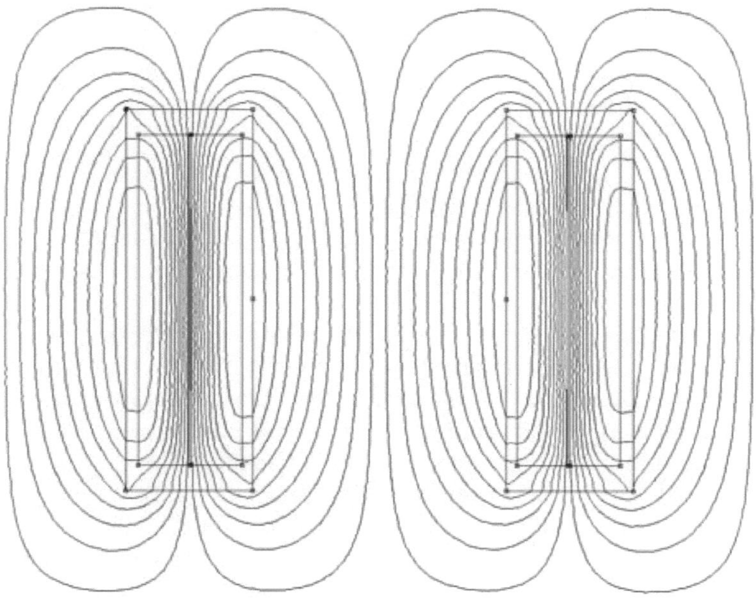

Figure 66.
Magnetic flux distribution in transformer under the short circuit condition.

These values are introduced to the *preprocessor* in the corresponding coils. It is recommended to refine the mesh and to increase the number of finite elements around the windings and in the air because there is a main part of the leakage flux. Then calculation and analysis of the results can be made. In **Figure 66**, the distribution of the flux lines, which correspond to the short circuit condition, is seen. As it was supposed, the flux lines are in the surrounding of the coils because the main flux is neglected. In this condition, no saturation occurs; therefore the calculation of the leakage inductance can be made by means of the energy of electromagnetic field and for comparison also from the linkage magnetic flux. It is recommended to do it by both ways.

The first way starts at setting of the z-coordinate on the value l_{av} = 161 mm. Then the calculation is launched to calculate the total leakage inductance. At the calculation based on the magnetic field energy, the whole cross-section area of transformer is marked in *postprocessor*. Then it is valid that:

$$L_{\sigma1} + L_{\sigma2}' = L_{\sigma} = \frac{2W}{I_{1N}{}^2} = \frac{2 \cdot 0.0193107}{2.75^2} = 5.1\ \text{mH}$$

or based on the linkage magnetic flux, but in *postprocessor* only the areas corresponding primary and secondary windings are marked and calculate A.J. Then it is valid that:

$$L_{\sigma1} + L_{\sigma2}' = L_{\sigma} = \frac{\psi}{I_{1N}^2} = \frac{\oint \boldsymbol{A} \cdot J\text{d}V}{I_{1N}^2} = \frac{0.0386215}{2.75^2} = 5.1\ \text{mH}$$

The second way is based on the fact that *z*-coordinate is set on 1 mm. Therefore, the value of energy must be multiplied by the average length of the conductor l_{av} = 161 mm. For the calculation from the magnetic field energy, it is valid that:

$$L_{\sigma1} + L_{\sigma2}' = L_{\sigma} = l_{av}\frac{2W}{I_{1N}{}^2} = 161\frac{2 \cdot 0.000119942}{2.75^2} = 5.1\text{mH}$$

or from the linkage magnetic flux by means of integral A.J:

$$L_{\sigma 1} + L_{\sigma 2}' = L_{\sigma} = l_{av}\frac{\psi}{I_{1N}^2} = l_{av}\frac{\oint \mathbf{A} \cdot J dV}{I_{1N}^2} = 161\frac{0.000239885}{2.75^2} = 5.1 \text{ m}H$$

The measured value of total leakage inductance is 5.2 mH (see **Table 7**), which means a very good coincidence of the results.

In this FEMM program, it is possible to calculate approximately the resistance of primary and secondary winding. For more precise calculation, the 3D program would be needed.

The resistance depends on the electrical conductivity of copper from which the windings are made. In simulating the value of copper, specific electrical conductivity σ = 58 MS/m is used or can be set based on the material library in the FEMM program. The calculation starts from the short circuit simulation, whereby z-coordinate is set on 1 mm. Now the average length of the turn of primary l_{avp} and secondary l_{avs} winding must be calculated. The calculation is made based on the geometrical dimensions in **Figures 56** and **65**. Then l_{avp} = 255.2 mm and l_{avs} = 339.2 mm. After the calculation, in *postprocessor*, the blocks must be marked, which correspond to the primary winding, and by means of the command *Resistive losses*, the Joule losses in the primary winding ΔP_j are calculated. In the same way, the losses in the secondary winding are calculated. Nevertheless, it is the same value because the cross-section area of the winding and current density is the same. The resistance is then calculated at 20°C for both windings as follows:

$$R_p = l_{avp}\frac{\Delta P_j}{I_{1N}^2} = 255.2\frac{0.050276}{2.75^2} = 1.71\ \Omega$$

$$R_s = l_{avs}\frac{\Delta P_j}{I_{2N}^2} = 339.2\frac{0.050276}{26.3^2} = 24.6\ \text{m}\Omega$$

For comparison the measured values are R_p = 1.91 Ω and R_s = 20 mΩ.

In the end the simulated and measured values of equivalent circuit parameters are summarized in **Table 8**. It can be proclaimed that the values obtained by simulation and measurement are in good coincidence.

	Measurement	FEMM	Deviation
Magnetizing inductance L_μ [H]	2.82	2.85	1.05%
Total leakage inductance L_σ [mH]	5.2	5.1	1.9%
Primary winding resistance R_p [Ω]	1.91	1.71	10.4%
Secondary winding resistance R_s [Ω]	0.02	0.0246	18.6%

Table 8.
Comparison of the equivalent circuit parameters.

3.5 Analysis of asynchronous machine parameters

Asynchronous motor parameters are simulated based on the no load test and locked rotor test in accordance with the equivalent circuit parameters [21]. Also calculation of the air gap electromagnetic torque and its ripple is made. Analysis is made for a real three-phase squirrel-cage asynchronous motor (its type symbol is 4AP90L); the nameplate and rated values are in **Table 9**. A picture of its geometrical parts and their dimensions are in **Figures 67** and **68**.

Rated stator voltage U_{1N}	400 V
Stator winding connection	Y
Rated power P_N	1500 W
Rated frequency f	50 Hz
Rated speed n	1410 min^{-1}
Phase number m	3
Rated slip s_N	6%
Number of pole pairs p	2
Rated torque T_N	10.15 Nm
Number of one-phase turns N_s	282
Number of slots per phase per pole q	3
Winding factor k_w	0.959
Active length of the rotor l_{Fe}	98 mm
Number of conductors in the slot z_Q	47
Magnetizing current I_0 obtained from no load measurement	2.3 A
L_μ magnetizing inductance obtained from no load measurement	0.32 H
Rated stator current I_{sN}	3.4 A

Table 9.
Nameplate and parameters of the investigated three-phase asynchronous motor.

3.5.1 Simulation of the no load condition

An ideal no load condition is defined at synchronous speed of the rotor, when rotor frequency is zero. In fact, at real no load condition, the rotor rotates at speed lower than synchronous speed, but the difference is not significant. Therefore, an ideal no load condition can be investigated without a big error. A magnetizing inductance of the equivalent circuit and also fundamental harmonic of the magnetic flux density in the air gap can be calculated by means of FEM. Here is a procedure how to do it:

- Draw a model of the investigated motor in a cross-section area in a program of FEMM (**Figure 69**).
- Enter the three-phase currents to the stator windings, materials, boundary conditions, and a mesh density. (The rotor currents are in the ideal no load condition zero).
- After the calculation, analyze in *postprocessor* distribution of the air gap magnetic flux density
- To make a Fourier series of the air gap magnetic flux density, calculate its fundamental harmonic, electromotive force (induced voltage), and from it the magnetizing inductance.

3.5.1.1 Drawing of the asynchronous motor model

From **Figure 69** it is seen that the asynchronous motor cross-section area is much more complicated than the transformer. There are more possibilities how to

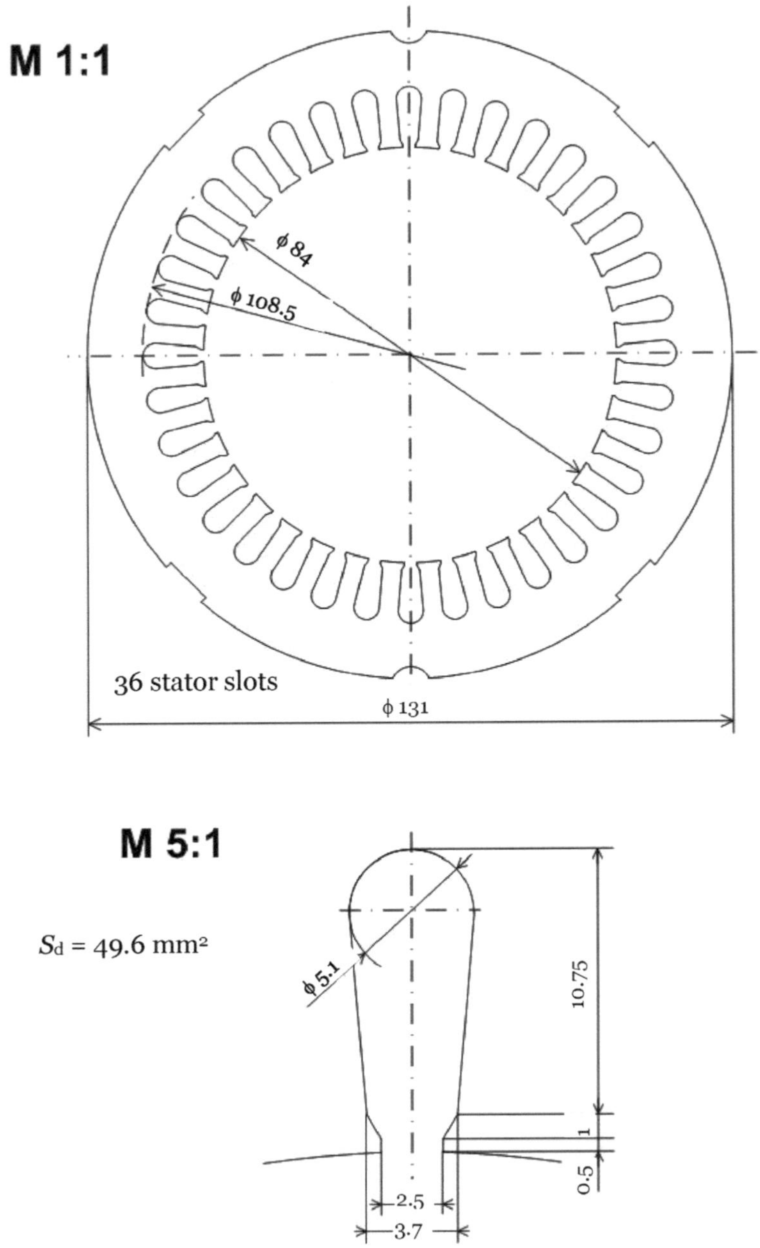

Figure 67.
Cross-section area of the stator sheet and detail of the stator slot with its geometrical dimensions.

draw this model, either directly in FEMM or in other graphical program (e.g., AUTOCAD, CAD, with the suffix *.dxf), and then to import it into FEMM. Here a drawing of the editor of the FEMM program is explained.

The first step is the setting of the task type which is investigated. In the beginning, when a new problem is investigated, the program calls up the user to define the type of the problem. For the no load condition, the *Magnetostatic Problem* is set. In the block *problem*, the units of the geometrical dimensions are set, usually in mm. Stator and rotor frequency is zero, because only one instant is investigated. In the block *Depth*, the active length of iron l_{Fe} is set. *Problem Type* is planar (see **Figure 69**).

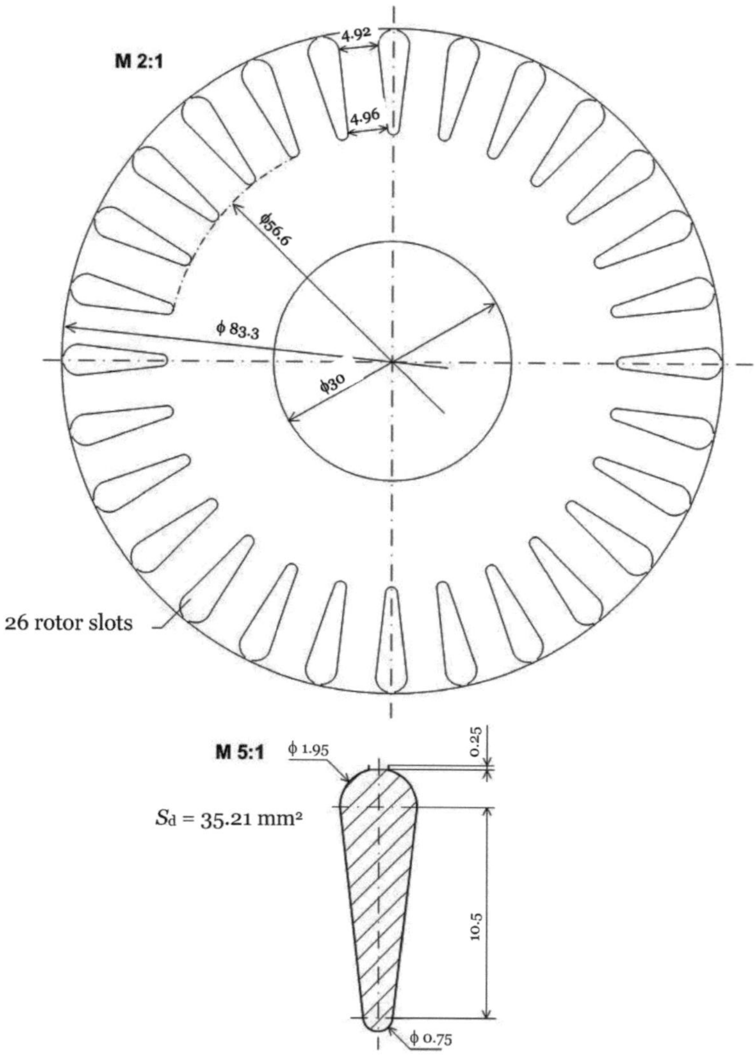

Figure 68.
Cross-section area of the rotor sheet and detail of the rotor slot with its geometrical dimensions.

The drawing starts with changing over the drawing editor to the *point mode*, and by means of tabulator, the points based on the *x*- and *y*-coordinates are set. It is recommended to draw the model in such a way that the center of the machine has coordinates 0,0. After the points are drawn, change the program into the *line mode* or *arc mode*, and join the points by straight lines or curves. If there is curve mode, the user is asked, which angle should have the curve, e.g., for semicircle it is 180°, and what the accuracy should be. According to the accuracy of the calculation, it is recommended to enter the number 1, higher number means lower accuracy.

If the same objects are drawn several times, e.g., stator or rotor slots, it is possible to apply copying, which is in the block *Edit* and *Copy*.

The next step is the setting of the materials and currents into the appropriate blocks. All bordered areas created during the drawing present the blocks into which it is necessary to input the materials. On the toolbar, it is necessary to change over to *block label, group mode*, to mark all blocks, and to define them. But before that, the materials must be designated and defined.

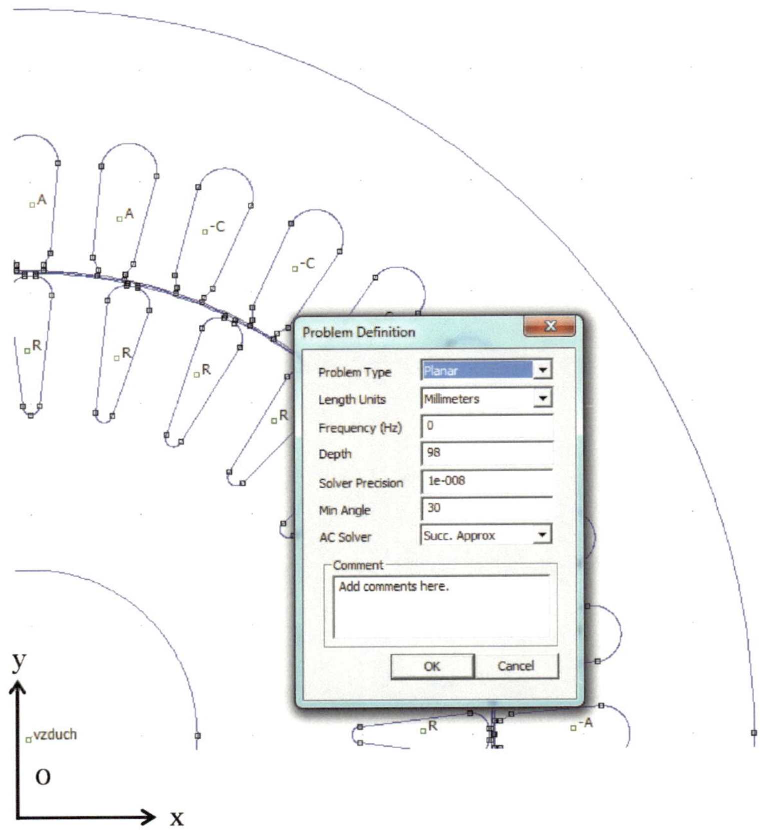

Figure 69.
The 1/4 of the cross-section area of the squirrel-cage asynchronous motor together with basic settings.

Materials are defined in *Properties – Add properties*, where the constants for materials are entered (**Figure 70**). In asynchronous machine, there is air in the air gap and ventilating channels; ferromagnetic circuit, which is defined by a nonlinear magnetizing B-H curve (in this case the employed sheet is Ei70, the thickness of the sheet is 0.5 mm, block name in the program is core); and material, from which the stator and rotor conductors are produced. It is usually copper or aluminum. These materials can be copied from the program library. For stator slots, the current density, corresponding to no load condition, must be entered. A calculation of the current density for stator slots is as follows:

Think one instant of the three-phase no load current for all three phases. For example, if phase A crosses to zero, then phases B and C have the values equaled to sin60° from the magnitude. If the next phase sequence is assumed around the stator circumference: +A, -C, +B, -A, +C, -B, etc., then A = 0, −A = 0, -C = $-J_{max}$ sin 60°, C = $+J_{max}$ sin 60°, B = $-J_{max}$ sin 60°, and −B = $+J_{max}$ sin 60°, where J_{max} is the magnitude of the current density. It can be calculated as follows:

$$J_{max} = \frac{z_Q I_{max}}{S_d} = \frac{47 \cdot \sqrt{2} \cdot 2.3}{49.6} = 3.082\ \mathrm{MA/m^2} \tag{482}$$

where I_{max} is the magnitude of the no load current, which flows along the conductors, z_Q is a number of the conductors in the slot, and S_d is a cross-section area of the stator slot (**Figure 67**). In calculation there is a no load current I_0 = 2.3 A used (see **Table 9**).

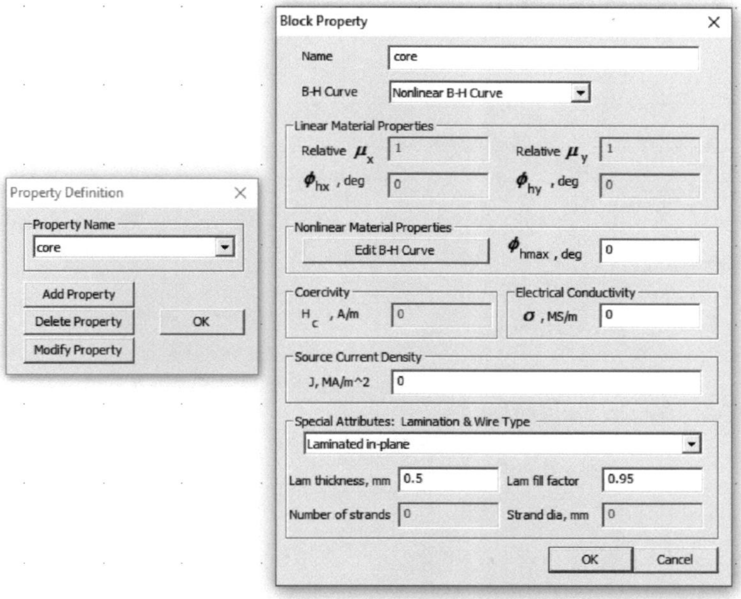

Figure 70.
Material designation and definition.

The calculated current densities are entered to the appropriate slots. In this case the number of slots per phase per pole is $q = 3$.

If all materials are defined, then it is necessary to allocate them to the appropriate blocks. The demanded block is designated by the right mouse button, and by pushing the space key, it is possible to allocate the material to the block. In this way, all blocks are defined. If it happens that a block is forgotten, it is not possible to make calculation until the block is not designated.

3.5.1.2 Setting of the boundary conditions

The last task before the calculation is launched, which is definition of the boundary conditions. Because the whole cross-section area of the asynchronous motor is analyzed, the Dirichlet boundary condition on the outer circumference, which is zero magnetic vector $\boldsymbol{A} = 0$, can be applied (**Figure 71**).

In the block *Properties*, click on the *Boundary* and define the boundary condition according to **Figure 71**. Then change over to the curve mode, choose the stator external circle by right mouse button, and by means of space key the appropriate boundary condition is allocated to the circle.

Before starting the calculation, click on the icon *Mesh*, and create the demanded mesh of the finite elements. If it looks in some parts to be widely spaced, it can be densified in the next way: It is necessary to change over to the block mode, by means of the right mouse button, choose the demanded block, mark *Let triangle choose Mesh size*, and set the required value of the finite elements. Then the very calculation can be started by means of the icon *solve*.

3.5.1.3 Calculation of the air gap magnetic flux density

After the calculation is finished, a distribution of the magnetic flux lines in the cross-section area of the investigated motor appears on the screen (see **Figure 72**). For illustration there is the whole cross-section area of the four-pole asynchronous motor.

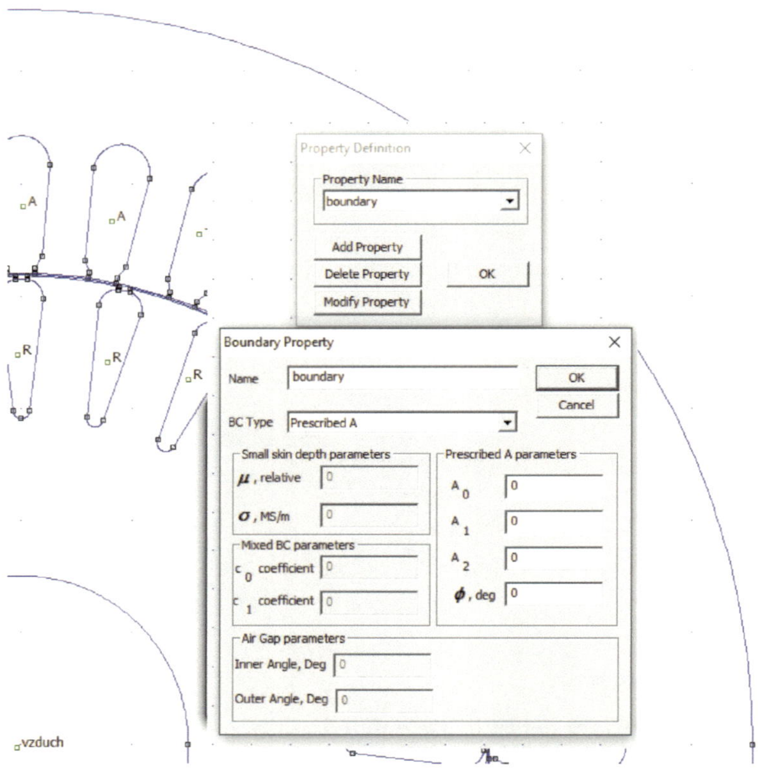

Figure 71.
Definition of the boundary conditions.

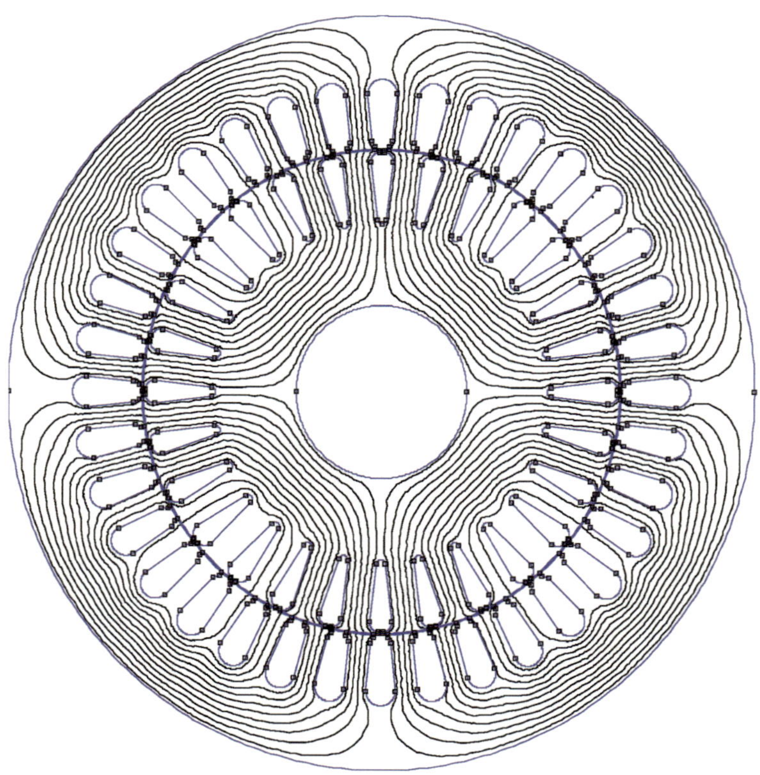

Figure 72.
Distribution of the magnetic flux lines in four-pole asynchronous motor in no load condition.

If a distribution of the magnetic flux density in the whole cross-section area is needed, it is possible to see its value in each point of the cross section and accord these values to optimize construction of the motor.

Most important is to know the shape of the waveform of the air gap magnetic flux density. Therefore, after the calculation in *postprocessor*, it is necessary to mark a circle in the middle of air gap (see **Figure 73a**) and on this circle in dialog window, to mark calculation of the normal component of the magnetic flux density. These values can be saved in data file and their elaboration in the other program continued. In such a way based on the Fourier series, the fundamental harmonic component can be calculated. In **Figure 73b**, a waveform of the air gap magnetic flux density is shown. After the Fourier series is made, it is possible to calculate electromotive force (induced voltage) and from this value the magnetizing inductance.

Based on the geometrical dimensions, a calculation of the induced voltage can be made. The induced voltage in the three-phase alternating rotating machines is given by an expression as follows:

$$U_i = \sqrt{2}\pi f \Phi_{\text{av}} N_s k_{\text{w}} = \sqrt{2}\pi f \frac{2}{\pi} B \frac{\pi D}{2p} l_{\text{Fe}} N_s k_{\text{w}} \tag{483}$$

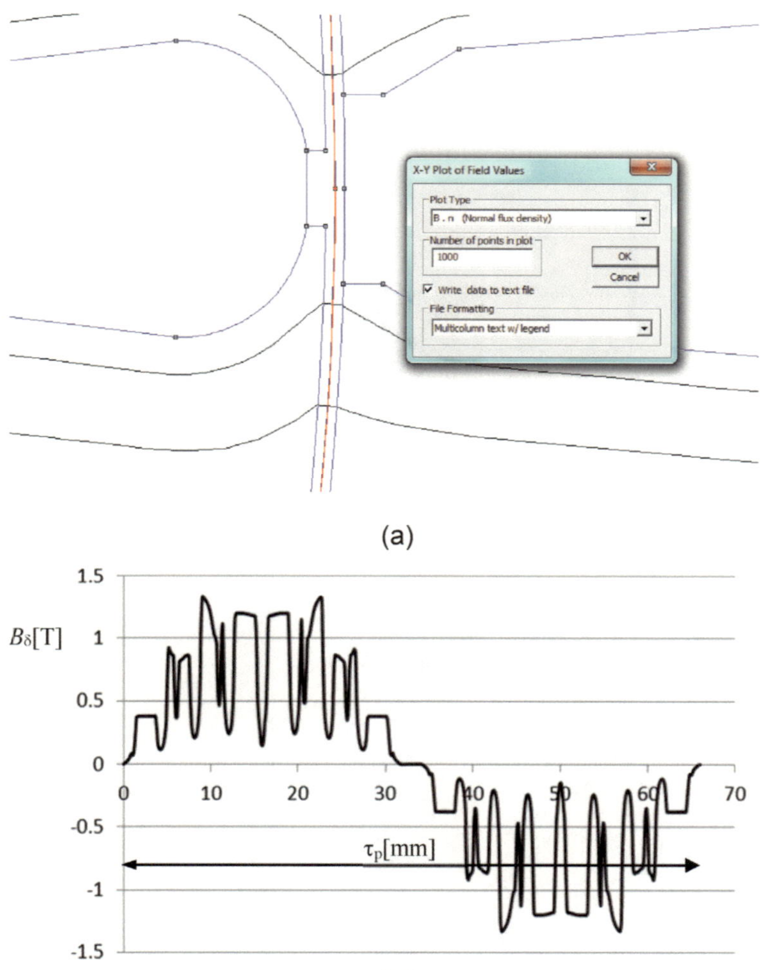

Figure 73.
(a) Marking of the circle in the middle of the air gap to calculate a normal component of the magnetic flux density and (b) distribution of the normal component of the air gap magnetic flux density in the investigated asynchronous motor.

where D = 84 mm, l_{Fe} = 98 mm, N_s = 282 turns, and k_w = 0.959. From the waveform of the air gap magnetic flux density in **Figure 73b**, the magnitudes of the harmonic components $B_{\delta\nu}$ can be calculated by means of the Fourier transformation. In **Figure 74**, the harmonic analysis (Fourier series) of the air gap magnetic flux density for 60 harmonic components is shown. To make a Fourier transformation, the program MATLAB can be employed.

The magnitude of the air gap magnetic flux density of the fundamental harmonics is $B_{\delta 1max}$ = 0.94 T. After this value is introduced to Eq. (483), the rms value of phase induced voltage fundamental harmonic is obtained:

$$U_{i1} = \sqrt{2}\pi f \frac{2}{\pi} B \frac{\pi D}{2p} l_{Fe} N_s k_w = \sqrt{2}\pi \frac{2}{\pi} 0.94 \frac{\pi \cdot 0.084}{4} 0.098 \cdot 282 \cdot 0.959 = 232.4 \text{ V}$$

The total voltage could be calculated by the sum of all harmonic components. The terminal voltage given in the nameplate is 3 x 400 V at star connection. Then the phase value is 230 V, which is in very good coincidence with the value obtained by FEMM. For illustration in **Figure 75**, there is a distribution of the magnetic flux density in the cross-section area of the investigated motor at no load condition.

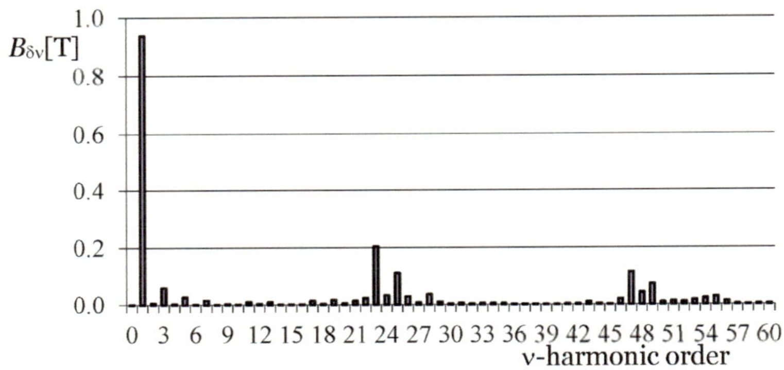

Figure 74.
Magnitude spectrum of the air gap magnetic flux density harmonic components in no load condition.

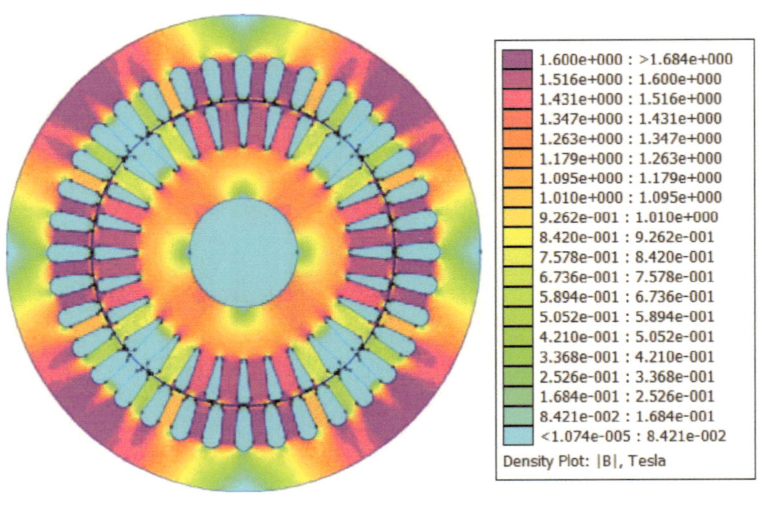

Figure 75.
Distribution of the magnetic flux density in no load condition.

3.5.1.4 Calculation of the magnetizing inductance at no load condition

At no load condition, also magnetizing inductance can be calculated. If the induced voltage, frequency, and no load current (from the no load measurement) are known, the magnetizing inductance is as follows:

$$L_{\mu} = \frac{U_{\mathrm{i}}}{\omega I_0} = \frac{232.4}{2 \cdot \pi \cdot 50 \cdot 2.3} = 0.321 \text{ H}$$

It was supposed that at no load condition, the magnetizing current is almost identical with no load current. The measured value of the magnetizing inductance is L_{μ} = 0.32 H (see **Table 9**), which is in very good coincidence of the results obtained from FEMM and measurement.

A calculation of magnetizing inductance can be done also based on the linkage magnetic flux and integral A.J, because in this no load condition, the ferromagnetic circuit is in the nonlinear area. A procedure can be described as follows: After the calculation in *postprocessor*, all slots corresponding with the stator winding are marked, and the integral A.J is calculated (**Figure 76**). It must be realized that the flux is created by all three phases; therefore the value of no load current is multiplied by 3, or linkage magnetic flux is divided by 3. Then the magnetizing inductance is:

$$L_{\mu} = \frac{\psi}{3I_0} = \frac{\oint \boldsymbol{A} \cdot J dV}{3I_0^2} = \frac{5.18345}{3 \cdot 2.3^2} = 0.326 \text{ H}$$

It is seen that a coincidence with the values obtained by other methods is very good.

3.5.2 Simulation of locked rotor condition

During the no load simulation by means of the FEMM, a magnetostatic field was employed. It is possible to do it also during the simulation of the locked rotor

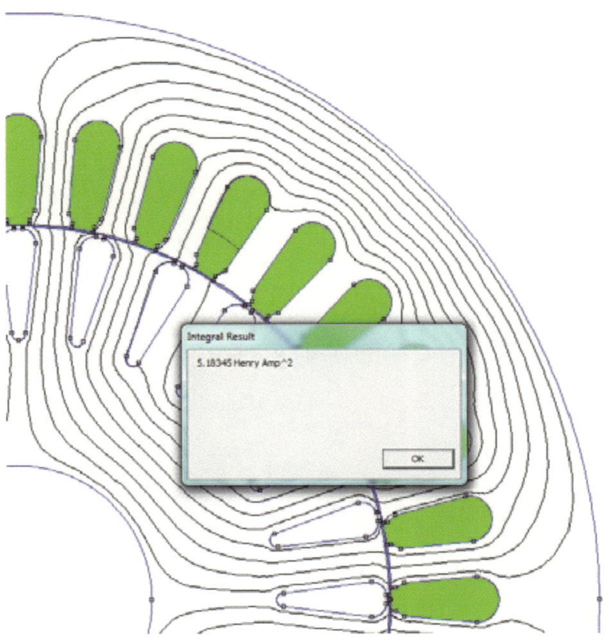

Figure 76.
Marking of stator slots for magnetizing inductance calculation.

condition, but it is necessary to enter to the rotor bars actual rotor currents, induced there from the stator currents. Such setting of the rotor currents is time-consuming, mainly if there is bigger number of the rotor slots. For completeness this way is described in Section 5.7, where the rated condition and rated torque are calculated.

Simulation of the locked rotor condition means to solve a harmonic task. It means to enter the frequency corresponding with the actual rotor frequency at locked rotor condition. Because in FEMM program there is no possibility to enter more various frequencies in one solution, it is necessary to enter to the base settings (**Figure 69**) the slip frequency. At the locked rotor condition, the slip is equal to 1, and corresponding frequency is frequency of the stator current f = 50 Hz. During the analysis, the same cross-section area of the asynchronous motor as in the no load condition is employed. In stator slots, the material of the stator slots (Cu, Al) and its conductivity and permeability, which is close to the vacuum, is entered. Usually the stator conductors in asynchronous motors are copper (copper electrical conductivity is 58 MS/m) and rotor cage is aluminum, the electrical conductivity of which is 24.59 MS/m. It is also the case of motor simulated here.

To define the currents, a block *Circuits* is employed. It is in the block below *Properties*, under *Property definition* and *Circuit Property* (see **Figure 77**).

In **Figure 78**, there are phasors of stator currents $\boldsymbol{I}_s$ for corresponding phases, eventually circuits. The circuits can be set on the *Parallel or Series* (**Figure 77**). Here the circuit *Series* is employed, and then the currents definition is in coincidence with **Figure 78**. It is a product of the complex figure which corresponds to the current phasor position, number of the conductors in the slot, and magnitude of the current. In this case a current is set on its rated value I_N = 3.4 A, and then its magnitude is I_{Nmax} = 4.8 A. The phases are excited as follows:

Phase A+: $(1+j0)z_Q I_{Nmax} = (1+j0)47\sqrt{2}\cdot 3.4 = (226+j0)$

Phase A−: $(-1+j0)z_Q I_{Nmax} = (-1+j0)47\sqrt{2}\cdot 3.4 = (-226+j0)$

Phase B+: $\left(-\frac{1}{2}-j\frac{\sqrt{3}}{2}\right)47\cdot\sqrt{2}\cdot 3.4 = (-113-j195.7)$

Phase B−: $\left(\frac{1}{2}+j\frac{\sqrt{3}}{2}\right)47\cdot\sqrt{2}\cdot 3.4 = (113+j195.7)$

Phase C+: $\left(-\frac{1}{2}+j\frac{\sqrt{3}}{2}\right)47\cdot\sqrt{2}\cdot 3.4 = (-113+j195.7)$

Phase C−: $\left(\frac{1}{2}-j\frac{\sqrt{3}}{2}\right)47\cdot\sqrt{2}\cdot 3.4 = (113-j195.7)$

From the winding theory, it is known:

$$q = \frac{Q_S}{2pm} = \frac{36}{2\cdot 2\cdot 3} = 3 \qquad (484)$$

where q is number of slots per phase per pole and p is number of pole pairs.

It means that the three adjacent stator slots are defined by the same phase current,

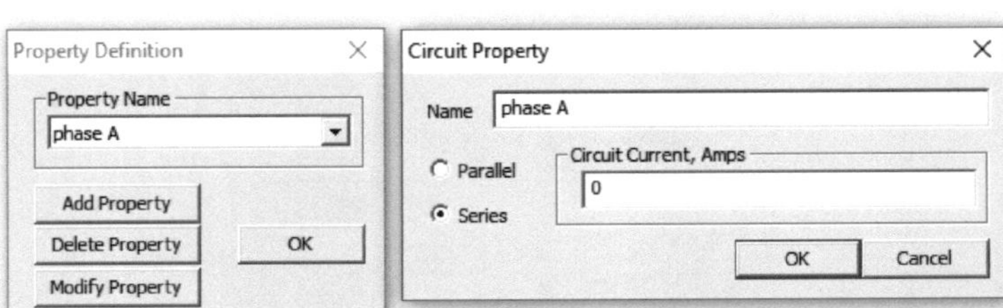

Figure 77.
Currents definition in locked rotor condition.

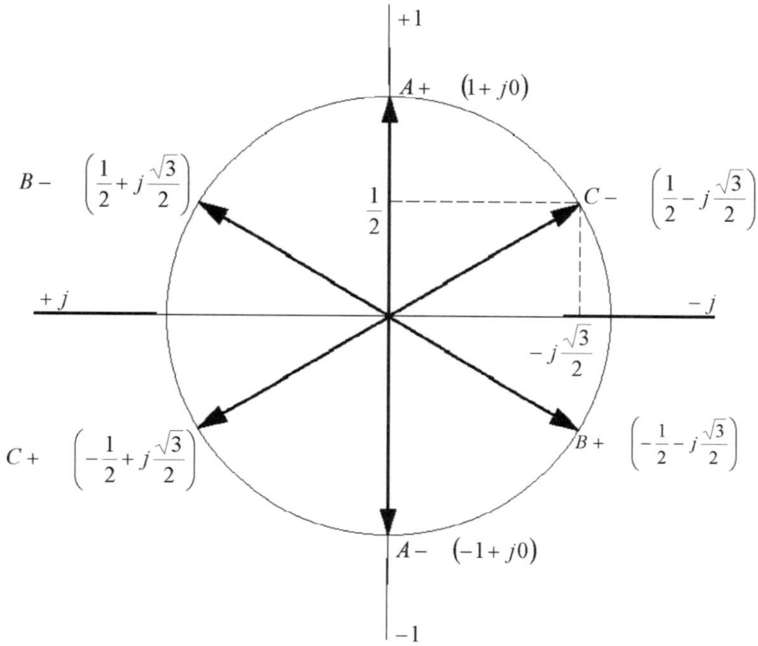

Figure 78.
The current phasors in complex plain for the locked rotor condition.

e.g., A+. These triplets of the slots are changed in the order, A+, A+, A+, C-, C-, C-, B+, B+, B+, A-, A-, A-, C+, C+, C+, B-, B-, B-, which is seen also in the **Figure 78**, where this changing of the phases creates clockwise rotating magnetic field.

As the number of the pole pairs $p = 2$, this order is repeated two times around the stator inner circumference. If there is double-layer winding, it must be taken into account. In here investigated motor, there is a single-layer winding.

In rotor conductors, the aluminum with its electrical conductivity is set, because at the harmonic task the currents are induced in the rotor conductors.

3.5.2.1 Calculation of the equivalent circuit parameters from the locked rotor simulation

The equivalent circuit parameters needed for the calculation of its parameters in locked rotor condition are in **Figure 79**. The resistance and leakage inductance

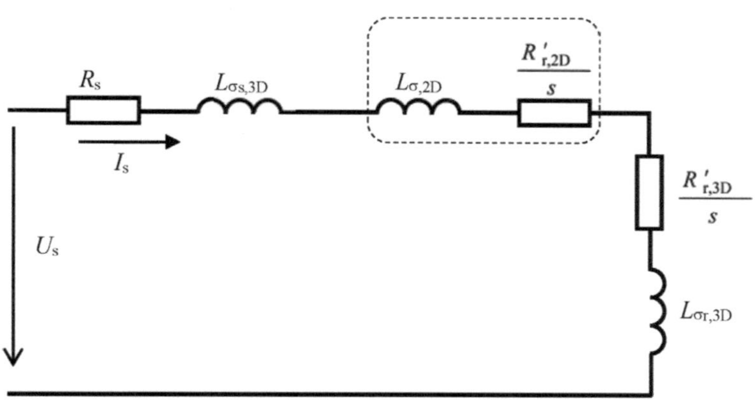

Figure 79.
Modified equivalent circuit of the asynchronous motor.

depend also on the frequency. In locked rotor condition, the rotor frequency is identical with the stator frequency; therefore in starting settings of the FEMM program, (*problem*) f = 50 Hz is entered.

In the 2D program, it is not possible to calculate resistance and leakage inductance of the end connectors and rotor end rings. For a correct calculation, these parameters must be calculated in another way or employ a 3D program.
In **Figure 79** there is an adapted equivalent circuit, where the dashed line shows which parameters can be calculated by 2D program.

Parameters out of dashed line are caused by 3D effect. They are the following: R_s, stator winding resistance; $L_{\sigma s,3D}$, stator leakage inductance of end windings; $L_{\sigma,2D}$, stator and rotor leakage inductance without end windings and rotor rings; $\frac{R'_{r,2D}}{s}$, rotor resistance referred to the stator without rotor rings; $\frac{R'_{r,3D}}{s}$, resistance of rotor rings referred to the stator; and $L_{\sigma r,3D}$, leakage inductance of the rotor rings. Supplied current in the simulation is rated current which corresponds also to locked rotor measurement. Magnetic flux lines distribution in locked rotor state is shown in **Figure 80**. The following calculations are carried out in accordance with [11].

During locked rotor simulation, the following parameters can be obtained: $L_{\sigma,2D}$ and $R'_{r,2D}$. Usually, locked rotor test is done with low supply voltage, so no saturation effect is present. Then, leakage inductance $L_{\sigma,2D}$ can be calculated from stored energy:

$$L_{\sigma,2D} = \frac{2W}{3I_{sN}^2} = \frac{2 \cdot 0.367}{3 \cdot 3.4^2} = 21.1 \text{ mH} \tag{485}$$

From (485), it can be seen that the current is multiplied by 3. It is caused by three-phase supplying and all three phases create energy of magnetic field.

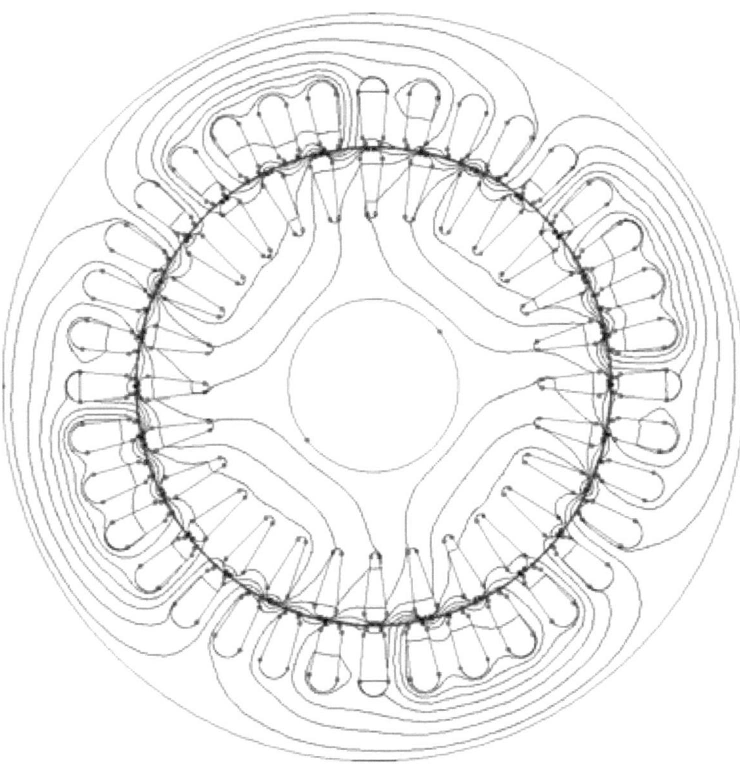

Figure 80.
Distribution of the magnetic flux lines of the asynchronous motor in the locked rotor condition.

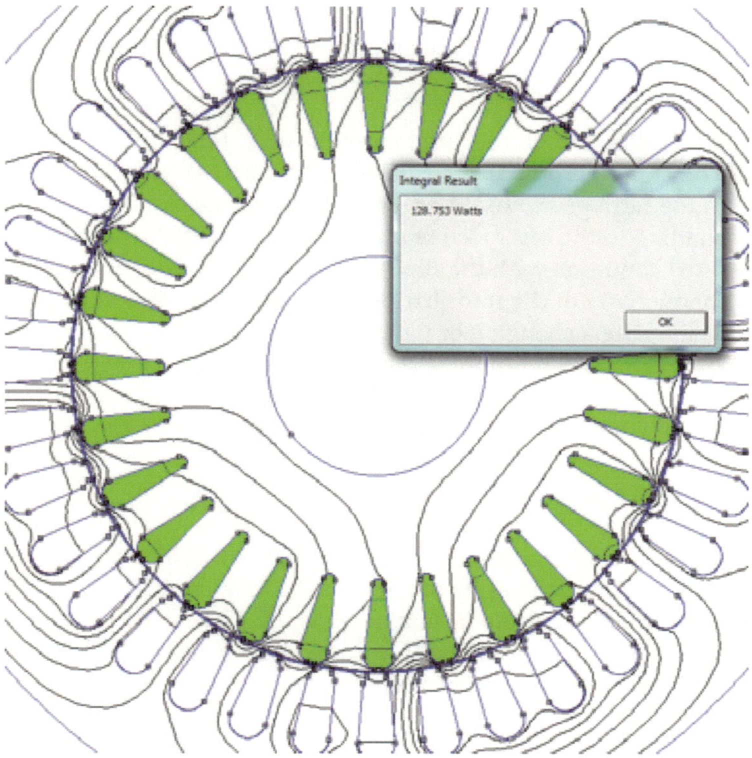

Figure 81.
Calculation of the losses in rotor bars.

The value of energy W can be obtained from whole cross-section area of the motor in *postprocessor*. To obtain total leakage inductance of the stator and rotor, analytical calculation of stator end winding leakage inductance and rotor ring leakage inductance must be taken into account, e.g., from [20], which is 26.8 mH. The value obtained from the locked rotor measurement is 29.5 mH, which is appropriate coincidence of the results.

The value of the rotor resistance is calculated from the losses in the rotor bars. In *postprocessor*, all blocks belonging to the rotor bars are marked (**Figure 81**) and by means of the integral, the ΔP_{jr} are calculated. Then the resistance of the rotor bars without the end rings is calculated as follows:

$$R'_{\mathrm{r,2D}} = \frac{\Delta P_{\mathrm{jr}}}{3 I_{\mathrm{sN}}^2} = \frac{128.755}{3 \cdot 3.4^2} = 3.71\ \Omega \tag{486}$$

The resistance of the rotor end rings is calculated from the expressions known from the design of electrical machines according to [1]. The total rotor resistance, which includes a bar and corresponding part of the end rings, referred to the stator side is 3.812 Ω. The measured value from the locked rotor test is 3.75 Ω, which is an appropriate coincidence of the parameters.

3.5.3 Calculation of the rated torque

The rated condition can be analyzed into two ways:

1. Magnetostatic task, at which instantaneous values of current densities are entered to the stator and rotor slots, which correspond to the same time instant

at investigated load, without a frequency. Then an electromagnetic torque is calculated around the circle in the middle of the air gap, according to Eq. (473).

2. Harmonic task, at which the rated currents are entered only to the stator slots. The currents in the rotor are calculated. There are two possibilities how to do it: (a) In the settings of FEMM program *problem*, the frequency is set, corresponding to the rotor frequency at the investigated load. For example, at the rated condition with the slip s_N = 6%, then the frequency f = 3 Hz is set. (b) A conductivity of the rotor bar proportional to the slip of investigated load is set, to simulate a changing of the current following the load. For example, if the rotor bars are aluminum, the electrical conductivity is σ_{Al} = 24.59 MS/m at the slip 1. At the rated load (s_N = 0.06), the conductivity is $s\sigma_{Al}$ = 0.06·24.59 = 1.475 MS/m, which corresponds to the lower current than for the slip 1.

3.5.3.1 Magnetostatic task

The procedure is similar as for the no load condition. It means to the stator slots the current densities corresponding to the phases and to the instantaneous values of the rated current are entered. An instant at which phase A crosses zero is taken into account. At this instant, phases B and C have an equal value but with opposite polarity. Then the current density in a slot belonging to the particular phase is as follows:

Phase A+: $J_A = \frac{I_N z_Q \sqrt{2}}{S_d}$, seeing that the current in phase A crosses zero, then J_A = 0.

Phase A−: $J_A = \frac{I_N z_Q \sqrt{2}}{S_d}$, seeing that the current in phase A crosses zero, then - J_A = 0.

Phase B+: $J_B = -\frac{I_N z_Q \sqrt{2}}{S_d} \sin 60^o$, because the phases are shifted about 120°, instantaneous value of the current density in phase B+ is negative.

Phase B−: $J_B = \frac{I_N z_Q \sqrt{2}}{S_d} \sin 60^o$, because the phases are shifted about 120°, instantaneous value of the current density in phase B- is positive.

Phase C+: $J_C = \frac{I_N z_Q \sqrt{2}}{S_d} \sin 60^o$, because the phases are shifted about 120°, instantaneous value of the current density in phase C+ is positive.

Phase C−: $J_C = -\frac{I_N z_Q \sqrt{2}}{S_d} \sin 60^o$, because the phases are shifted about 120°, instantaneous value of the current density in phase C- is negative.

The J is current density, I_N is rated current, z_Q is number of conductors in the slot, and S_d is slot cross-section area.

In the magnetostatic task, the currents must be entered also to the rotor, to be able to calculate the torque in the air gap. The current is referred from the stator to the rotor side and to the corresponding current density at particular time instant. The expression is known from the electrical machine design theory [1]:

$$J_r = I_N \frac{2mN_s k_w}{Q_r S_d} \sin \alpha \tag{487}$$

where m is stator phase number, N_s is number of turns of the stator phase, k_w is its winding factor, and Q_r is number of rotor bars. The number of the rotor turns is ½; S_d is cross-section area of the rotor slot. An angle α represents angular rotation of the rotor bar currents, and $\sin(\alpha)$ corresponds to the instantaneous value of the current density in each rotor slot. The angle α is calculated as follows:

$$\alpha = \frac{2p180}{Q_r} n \quad (488)$$

where n is a number of the rotor bar. Numbering of n is started from zero in that rotor slot, where current density starts from zero. After all current densities are entered, the calculation is launched. After the calculation is carried out, and the magnetic flux lines are depicted, the value of electromagnetic torque in the air gap can be calculated as follows:

1. In *postprocessor* a circle in the middle of the air gap is marked by a red line.

2. The value of the torque is calculated based on the Maxwell stress tensor (Section 3.2, according to **Figure 82**).

The obtained value of the electromagnetic torque at the rated current 3.4 A is 9.89 Nm (**Figure 82a**). For comparison, the rated torque on the shaft is T_N = 10.15 Nm and the value of the loss torque, obtained from the no load test, is T_{loss} = 0.7 Nm. Then the value of developed electromagnetic torque is:

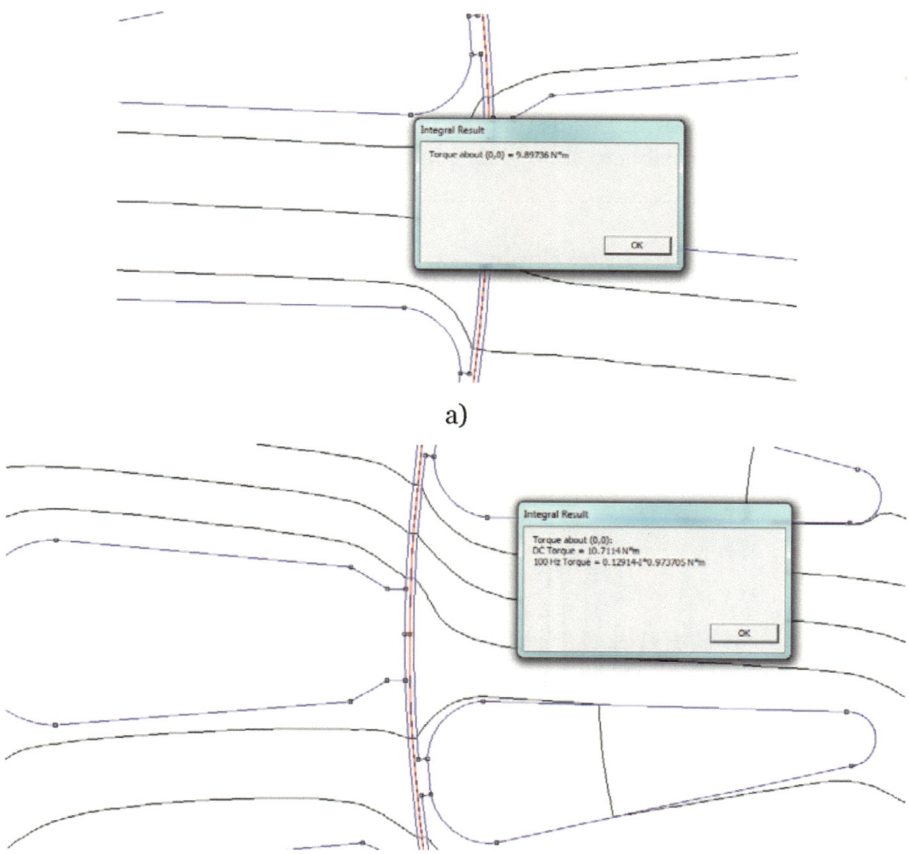

a)

b)

Figure 82.
Illustration figure for electromagnetic torque calculation based on the (a) magnetostatic task and (b) harmonic task.

$$T_e = T_N + T_{loss} = 10.15 + 0.7 = 10.85 \text{ Nm},$$

which is an appropriate coincidence of the results. The calculation can be done also based on the harmonic task.

3.5.3.2 Harmonic task

The other way how to analyze the rated condition of the asynchronous motor is calculation by means of harmonic task. The entered frequency is slip frequency, e.g., if the frequency of the stator current is 50 Hz and the rated slip is 6%, then the slip frequency is 50·0.06 = 3 Hz, which is employed for setting. In this case the calculated electromagnetic torque in the air gap is 10.69 Nm, which is better coincidence in comparison with calculation of magnetostatic task.

The developed electromagnetic torque can be calculated in such a way, that electrical conductivity proportional to the slip is entered to the rotor slots. The feeding frequency in the stator is 50 Hz. It is supposed that the electrical aluminum conductivity is σ_{Al} = 24.59 MS/m. It means at locked rotor condition, at slip equal 1, the conductivity is $s\sigma_{Al}$ = 1 24.59 = 24.59 MS/m and for the rated condition is $s_N\sigma_{Al}$ = 0.06·24.59 = 1.47 MS/m. In this case the calculated electromagnetic torque in the air gap is 10.71 Nm (**Figure 82b**), which is very close to the measured value. For the calculation of the electromagnetic torque, it is recommended to employ the harmonic task, from the point of view of accuracy and time demanding.

By means of the harmonic task, it is possible to calculate also other conditions of the asynchronous motor along the mechanical characteristic. It is possible also to parameterize the investigated problem and to program by means of LUA script in the FEMM program [10]. It means that the whole model is assigned in parameters mode and then it is possible to change its geometrical dimensions and optimize its properties.

3.5.4 Calculation of the torque ripple

The torque ripple is caused by the stator and rotor slotting and mainly by the stator and rotor slot openings. The calculation of the torque ripple is made in the air gap. A procedure is similar like in the case of the harmonic task during the calculation of the torque in rated condition. The stator slots are fed by rated current in coincidence with **Figure 78** (Section 5.5), and electrical conductivity of the rotor

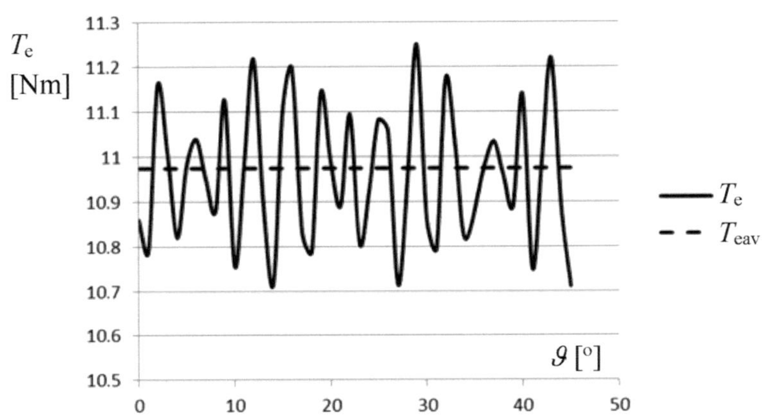

Figure 83.
Electromagnetic torque of the asynchronous motor in the air gap vs. rotor position, calculated based on the harmonic task.

slots is proportional to the rated slip $s_N\sigma_{Al}$ = 0.06·24.59 = 1.47 MS/m. The rotor position is changed gradually by 1° mechanical. The calculation was carried out for positions from 0° till 45° mechanical, which means one half of the pole pitch $\tau_p/2$. The calculated torque values vs. rotor position ϑ are shown in **Figure 83**. The average value of all calculated values is:

$$T_{eav} = \int_{\vartheta} T_e d\vartheta = 10.975\ \text{Nm}. \tag{489}$$

The torque ripple in the air gap presented in percentage is:

$$T_{ripp} = \frac{T_{emax} - T_{emin}}{T_{eav}} 100[\%] \tag{490}$$

From **Figure 83** it is seen that the ripple torque is moving in the interval from 1.45% till 4.88%.

3.6 Analysis of the synchronous machine parameters

From the point of view of operation principles and construction, the synchronous machines can be divided into three basic groups:

1. Synchronous machine with wound field coils creating electromagnetic excitation, which can be designed with salient poles or with cylindrical rotor (Section 6.1).

2. Synchronous reluctance machine with salient poles on the rotor without any excitation (Section 6.2).

3. Synchronous machine with permanent magnets on the rotor, creating magnetic flux excitation. The permanent magnets can be embedded in the surface of the rotor, or on the surface of the rotor (Section 6.3).

Here, parameter investigation by means of FEM of all three types of the synchronous machines is presented in the chapters as it is written above.

3.6.1 Synchronous machine with wound field coils on the rotor salient poles

In synchronous machine with wound rotor, an analysis of the no load condition, short circuit condition, synchronous reactances in d- and q-axis, and air gap electromagnetic torque is carried out by means of the FEM. The nameplate and parameters of the investigated generator are shown in **Table 10**. Its basic geometrical dimensions are in **Figure 84**.

3.6.1.1 Simulation of the no load condition of synchronous machine

The goal of the simulation in no load condition is to calculate induced voltage in the no load condition if the field current needed for this voltage is known. This current can be calculated during the design procedure or by a measurement on the real machine in generating operation. Here the measured value I_{f0} = 4.6 A, at the induced rated voltage U_i = U_{phN} = 230 V, is used.

Rated voltage U_{1N}	400 V
Stator winding connection	Y
Rated power S_N	7500 VA
Rated frequency f	50 Hz
Rated speed n	1500 min^{-1}
Rated power factor $\cos\varphi$	1
Rated stator current I_N	10.8 A
Rated field voltage U_f	32 V
Rated field current I_f	7.4 A
Number of pole pairs p	2
Number of field coil turns N_f	265
Number of turns of single stator phase N_s	174
Number of slots per pole per phase q	3
Number of stator slots Q_s	36
Number of conductors in the slot z_Q	29
Cross-section area of the rotor winding S_f	0.00081 m^2
Winding factor k_w	0.9597
Active length of the rotor l_{Fe}	80.2 mm
Diameter of the boring D	225 mm
Air gap δ	2 mm
Magnetizing current measured from the rotor I_{f0}	4.6 A
Magnetizing current measured from the stator side I_μ	12 A
Factor $g = I_{f0}/I_\mu$	0.383
Leakage stator reactance (measured) $X_{\sigma s}$	1.4 Ω
Magnetizing reactance in d-axis (measured) $X_{\mu d}$	21.6 Ω
Magnetizing reactance in q-axis (measured) $X_{\mu q}$	6.27 Ω

Table 10.
Nameplate and parameters of the analyzed synchronous generator.

A procedure is the same as in the case of asynchronous machine. It is started with *preprocessor*, where it sets *magnetostatic analysis*, *planar problem*, and zero *frequency*, because only field winding is fed by DC. *Z*-coordinate depth represents active length of the rotor, l_{Fe} = 80.2 mm, obtained by measurement.

Then based on the geometrical dimensions, a cross-section area is drawn as it is in **Figures 84** and **85**. The names of materials are allocated to the blocks. In this condition, when the stator currents are zero (no load condition), the rotor is in random position. However, it is better to set it to the d-axis, which is used also at the investigation of the short circuit condition. This rotor setting is made in such a way that the rotor pole axis is in coincidence with those slots in the stator, in which it would be zero current during the short circuit test. In this case, they are the slots corresponding with phase A (**Figure 86**).

Stator and rotor sheets are marked as ferromagnetic material. The stator winding is in single layer, and then in the stator slots, there are gradually phases A+, C-, B+, A-, C+, B-, which are now zero currents. The slots are defined only by copper permeability and conductivity. In the middle of the air gap is drawn an auxiliary circle on which distribution of magnetic flux density and electromagnetic torque is

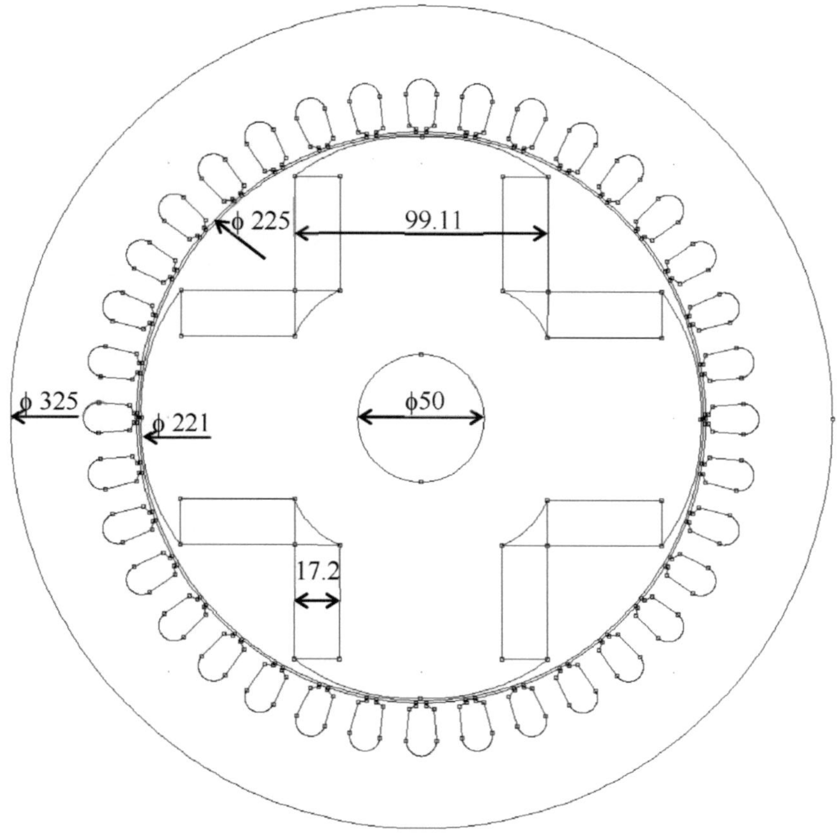

Figure 84.
Cross-section area of the investigated four-pole synchronous machine with salient poles and its geometrical dimensions.

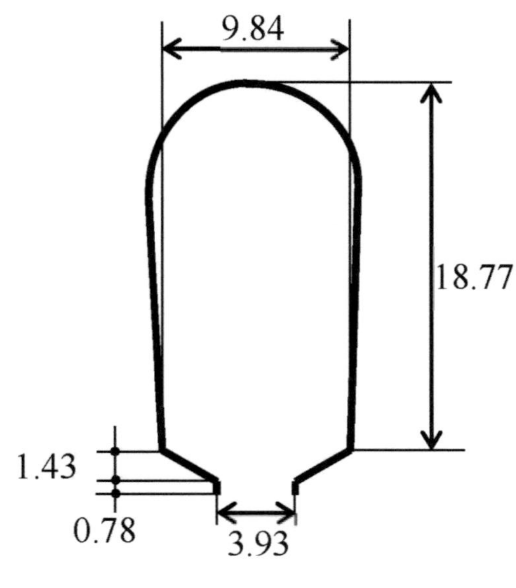

Figure 85.
Stator slot with its geometrical dimensions.

calculated. The field coils are marked as F1 and F2 (**Figure 86**), in which current density is set, corresponding with the current I_{f0}. The current density is calculated as follows:

$$J_f = \frac{N_f I_{f0}}{S_f} = \frac{265 \cdot 4.6}{0.00081} = 1.504938 \text{ MA/m}^2, \tag{491}$$

where S_f is cross-section area of the field coil on the single rotor pole, e.g., F1 **Figure 86**. With regard to the fact that the field current is DC, the current density in the block F1 is positive and in the block F2 negative. All defined materials and blocks are shown in **Figure 86**. Magnetizing characteristic is taken from the program library, because the real characteristic of the investigated machine is not known. Therefore, calculation for other field currents, as it is obvious during the measurement, is not made. If in other cases *B-H* characteristic is known, there is possibility to make calculations in its whole scope.

The boundary conditions are set in a similar way like in the case of asynchronous machine in the previous chapter. It is supposed that at zero magnetic vector potential, it means set A = 0. Then the mesh can be created and calculation launched. In the surroundings of the air gap, the mesh can be refined, to get more accurate results. After the calculation is finished, a distribution of the magnetic flux lines is described (**Figure 87a**). In *postprocessor* an auxiliary circle in the middle of the air

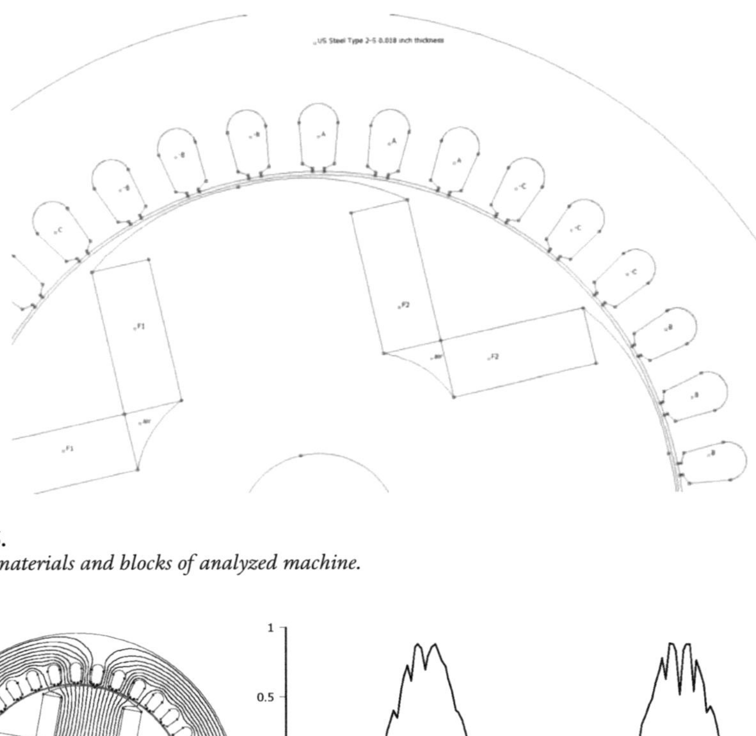

Figure 86.
Setting of materials and blocks of analyzed machine.

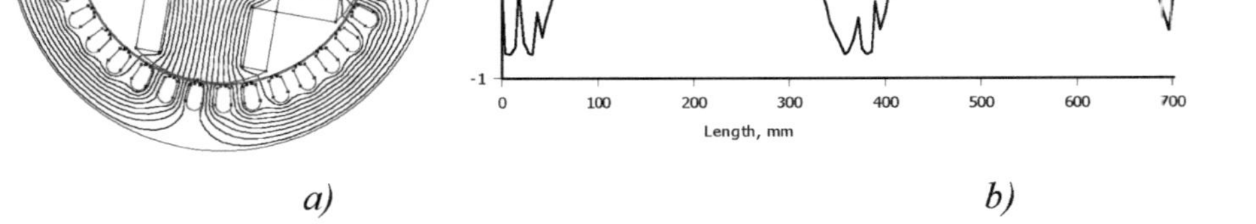

Figure 87.
Outputs obtained from postprocessor, (a) distribution of the magnetic flux lines in the no load condition and (b) magnetic flux density waveform along the whole air gap (by means of rotor field current I_{fo}).

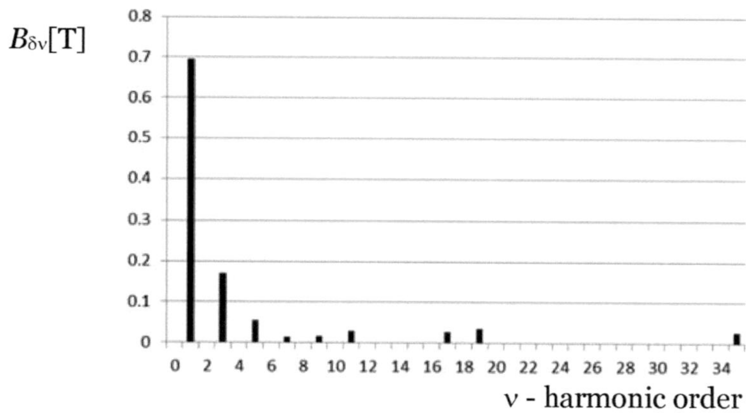

Figure 88.
Fourier series of air gap magnetic flux density in no load condition.

gap is marked and asked to show a waveform of the magnetic flux density around the whole circumstance (**Figure 87b**). Then harmonic components are calculated by means of Fourier series (**Figure 88**). The rms value of the induced voltage is calculated based on the magnitude of the fundamental harmonic component in the air gap $B_{\delta 1\text{max}}$.

Now it is possible to calculate a phase value of the fundamental harmonic of the induced voltage:

$$U_{\text{i}1} = \sqrt{2}\pi f \frac{2}{\pi} B_{\delta 1\,\text{max}} \frac{\pi D}{2p} l_{\text{Fe}} N_{\text{s}} k_{\text{w}} = \sqrt{2}\pi 50 \frac{2}{\pi} 0.69 \frac{\pi 0.225}{4} 0.08 \cdot 174 \cdot 0.959 = 230.78\ \text{V} \tag{492}$$

where $B_{\delta 1\text{max}}$ = 0.69 T (**Figure 88**). The calculated value of the induced voltage is in very good coincidence with the measured value, which is 230 V.

3.6.1.2 Induced voltage calculation by means of the stator current

If the field current is zero I_{f} = 0 A and the synchronous machine is applied to the grid with rated voltage U_{N}, then a magnetizing current I_{μ} flows in the stator winding to induce rated voltage equal to the terminal voltage. The same current flows in the stator winding if the terminals are short circuited and the machine is excited by the field current I_{f0}, in generating operation. The investigated synchronous machine was measured in short circuit condition in the generating mode, and the measured value is I_{μ} = 12 A.

To get the induced voltage on the terminals, the stator slots are fed by the current I_{μ} = 12 A at zero field current in the rotor. A magnetostatic analysis is chosen because only one time instant is analyzed. Settings in the blocks representing ferromagnetic materials are identical with the previous chapter. A difference is that the current density in the rotor is zero, and in the stator slots, there are current densities corresponding to the current I_{μ}. A calculation of the current density is made for the instant at which the current in phase A is zero. The instantaneous value of phases B and C has sin60° from the magnitude of the current I_{μ}, i.e., $I_{\mu\text{max}}$sin60°. If the sequence of the phases around the stator circumference is according to **Figure 86**, then:

Phase A: A+ = 0, A− = 0
Phase C: C− = $-J_{\mu\text{max}}$sin60°, C+ = + $J_{\mu\text{max}}$sin60°
Phase B: B+ = $-J_{\mu\text{max}}$sin60°, B− = + $J_{\mu\text{max}}$sin60°

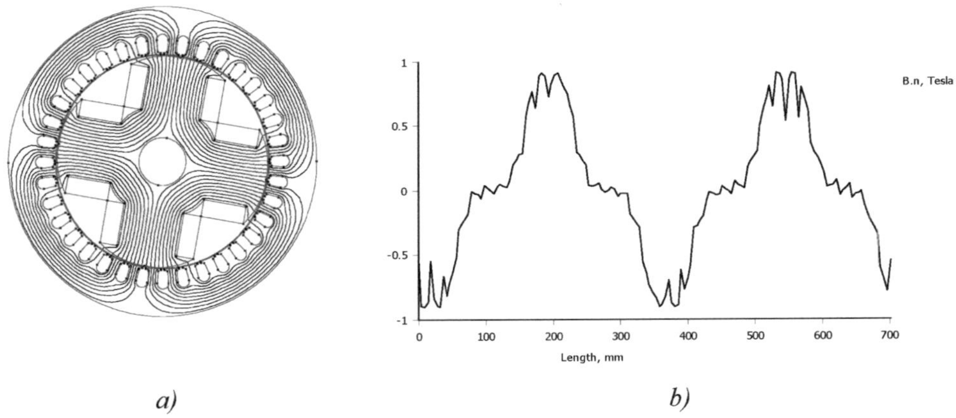

Figure 89.
Outputs from the postprocessor, (a) distribution of the magnetic flux lines if the stator slots are fed by the current $I_{\mu o}$ *and (b) waveform of the air gap magnetic flux density along the whole circumference of the air gap.*

where $J_{\mu\max}$ is magnitude of the current density. It is calculated as follows:

$$J_{\mu\max} = \frac{z_Q I_{\mu\max}}{S_d} = \frac{29 \cdot \sqrt{2} \cdot 12}{0.000207} = 2.377518 \text{ MA/m}^2 \qquad (493)$$

where z_Q is number of conductors in one slot and S_d is the area of the slot. In **Figure 86** there is a sequence of the slots and phases because the number of the slots per phase per pole $q = 3$. The calculated current densities are set to these slots. The boundary condition is the same as in the case of no load condition. In **Figure 89a** there is distribution of the magnetic flux lines, and in **Figure 89b** there is a waveform of the air gap magnetic flux density along the whole circuit. If these figures are compared with **Figure 87**, it can be seen that they are almost identical. Correctness of the simulation can be confirmed by the calculation of the induced voltage. The magnitude of the air gap fundamental harmonic component obtained based on the Fourier series is $B_{\delta 1\max} = 0.689$ T, which corresponds to the induced voltage $U_i = 230.45$ V, calculated by means of Eq. (492).

These results confirm that the induced voltage can be simulated by means of FEM and FEMM program very accurately by both ways.

3.6.1.3 Synchronous reactance X_d *and* X_q *calculation*

The synchronous reactance is a sum of the magnetizing reactance in d- or q-axis and stator leakage reactance $X_{\sigma s}$:

$$X_d = X_{\mu d} + X_{\sigma s} \qquad (494)$$

$$X_q = X_{\mu q} + X_{\sigma s}. \qquad (495)$$

The stator leakage reactance cannot be calculated by means of the 2D FEM, because the leakage flux of the stator end windings will not be included. It would be possible to do only in 3D FEM, which is not the case of the FEMM. In this chapter the magnetizing reactances in d- and q-axis are calculated by means of the FEMM, and the stator leakage reactance is taken from the measurement, $X_{\sigma s} = 1.4\ \Omega$.

Based on Ohm's law, it is valid that:

$$X_\mu = \frac{U_{i1}}{I_\mu} \qquad (496)$$

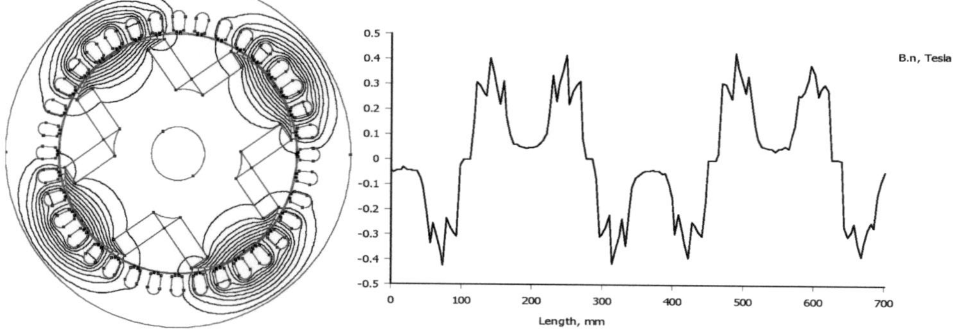

a) b)

Figure 90.
Outputs from the postprocessor, (a) distribution of the magnetic flux lines in q-axis, if the stator winding is fed by I_μ *= 12 A, and (b) waveform of the magnetic flux density along the air gap if the* $B_{\delta 1max}$ *is in the q-axis.*

This expression is valid in general; therefore it is used in d-axis and q-axis:

$$X_{\mu d} = \frac{U_{i1d}}{I_{\mu d}} \tag{497}$$

$$X_{\mu q} = \frac{U_{i1q}}{I_{\mu q}} \tag{498}$$

Equation (497) is valid if the rotor position is in d-axis and Eq. (498) if the rotor position is in q-axis. The machine is fed by three-phase magnetizing stator current I_μ. The rotor field current is zero. Because the calculation of the induced voltage in Section 3.6.2 was made for rotor position in d-axis, the results can be used now for the d-axis magnetizing reactance calculation. The values of the induced voltage are introduced to the expression (497) and then to the (494):

$$X_{\mu d} = \frac{U_{i1d}}{I_{\mu d}} = \frac{230.45}{12} = 19.2\ \Omega$$

$$X_d = X_{\mu d} + X_{\sigma s} = 19.2 + 1.4 = 20.6\ \Omega$$

The value obtained based on the slip method is X_d = 23 Ω, and the value obtained from the no load and short circuit measurement is X_d = 19.5 Ω, which shows an appropriate coincidence.

In the case of q-axis synchronous reactance calculation, it is necessary to move the rotor to the q-axis or to move the $B_{\delta 1max}$ to the q-axis. Here moving of the rotor to the q-axis is shown. The unaligned position of the rotor in q-axis (in the middle between two poles) is moved under the $B_{\delta 1max}$. The calculation is made at the same magnetizing current I_μ = 12 A as for the d-axis, because in this position a big air gap dominates, therefore saturation of the ferromagnetic circuit does not happen. It means that the proportion between the induced voltage and magnetizing current is constant (Eq. (498)). A distribution of the magnetic flux lines and waveform of the magnetic flux density in the air gap for the q-axis is seen in **Figure 90**.

Based on the Fourier series of the air gap magnetic flux density (**Figure 91**), the magnitude of the fundamental harmonic component is calculated: $B_{\delta 1max}$ = 0.22 T. The value of the induced voltage, if the rotor is positioned in the q-axis, can be calculated based on Eq. (492):

$$U_{\text{i1q}} = \sqrt{2}\pi f \frac{2}{\pi} B_{\delta 1\,\text{max}} \frac{\pi D}{2p} l_{\text{Fe}} N_{\text{s}} k_{\text{w}} = \sqrt{2}\pi 50 \frac{2}{\pi} 0.22 \frac{\pi 0.225}{4} 0.08 \cdot 174 \cdot 0.959 = 73.58 \text{ V}$$

The magnetizing reactance and synchronous reactance in q-axis are calculated by means of Eqs. (498) and (495):

$$X_{\mu\text{q}} = \frac{U_{\text{i1q}}}{I_{\mu\text{q}}} = \frac{73.58}{12} = 6.13 \ \Omega$$

$$X_{\text{q}} = X_{\mu\text{q}} + X_{\sigma\text{s}} = 6.13 + 1.4 = 7.53 \ \Omega$$

The value of the synchronous reactance in q-axis obtained based on the slip method measurement is X_q = 7.67 Ω. It is seen that there is again very good coincidence with measurement. Therefore, it can be proclaimed that if there are known geometrical dimensions of the machine and its *B-H* curve, it is possible to get very reliable results based on the FEM simulation.

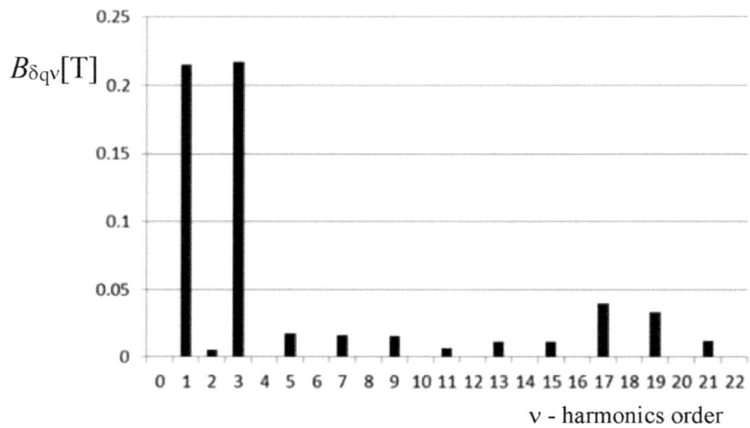

Figure 91.
Fourier series of the air gap magnetic flux density to determine the magnetizing reactance in q-axis.

3.6.1.4 Calculation of the synchronous machine electromagnetic torque

An electromagnetic torque in the air gap of the synchronous machine is calculated for the rated condition based on the nameplate data. Some parameters needed for simulation must be additionally calculated. The power of the investigated machine is 7.5 kW at the synchronous speed 1500 min^{-1} and cosφ = 1. Then it is possible to calculate the rated torque on the shaft:

$$T_{\text{N}} = \frac{P_{\text{N}}}{\Omega_{\text{N}}} = \frac{7500}{\frac{2\pi n}{60}} = \frac{7500}{\frac{2\pi 1500}{60}} = 47.74 \text{ Nm}. \quad (499)$$

The rated current in the stator slots is I_N = 10.8 A. It is needed to calculate its magnitude and current density according to Eq. (493):

$$J_{\text{Nmax}} = \frac{z_Q I_{\text{Nmax}}}{S_{\text{d}}} = \frac{29 \cdot \sqrt{2} \cdot 10.8}{0.000207} = 2.139766 \text{ MA/m}^2.$$

A magnetostatic case is chosen; therefore one time instant of the stator current is determined to be calculated. Most suitable is the instant at which phase A current crosses the zero value and the current densities of other phases are as follows:

$$J_{\mathrm{NB}} = -J_{\mathrm{Nmax}} \sin 60^o = -2.139766 \ \sin 60^o = -1.853091 \ \mathrm{MA/m^2}$$

$$-J_{\mathrm{NB}} = 1.853091 \ \mathrm{MA/m^2}$$

$$J_{\mathrm{NC}} = 1.853091 \ \mathrm{MA/m^2}$$

$$-J_{\mathrm{NC}} = -1.853091 \mathrm{MA/m^2}$$

$$-J_{\mathrm{NA}} = J_{\mathrm{NA}} = 0$$

The value of the field current I_{f}, current density, and load angle ϑ_{L} can be calculated from the phasor diagram according to **Figure 92**, where the stator winding resistance is neglected:

$$I_{\mathrm{f}} = \sqrt{(gI_{\mathrm{aN}})^2 + I_{\mathrm{f0}}^2} = \sqrt{(0.383 \cdot 10.8)^2 + 4.6^2} = 6.18 \ \mathrm{A}$$

Corresponding current density, according to Eq. (491), is:

$$J_{\mathrm{f}} = \frac{N_{\mathrm{f}} I_{\mathrm{f}}}{S_{\mathrm{f}}} = \frac{265 \cdot 6.18}{0.00081} = 2.021851 \ \mathrm{MA/m^2}$$

Now it is possible to calculate the load angle ϑ_{L}, which is important for moving the rotor in the FEMM program. Note that the phasor diagram is drawn for two-pole machine, i.e., in electrical degrees. Here the investigated machine is four poles, $2p = 4$; it means that in the FEMM program the moving of the rotor must be made in mechanical degrees and it is necessary to calculate it:

$$\vartheta_{\mathrm{Lmec}} = \vartheta_{\mathrm{L}}/p.$$

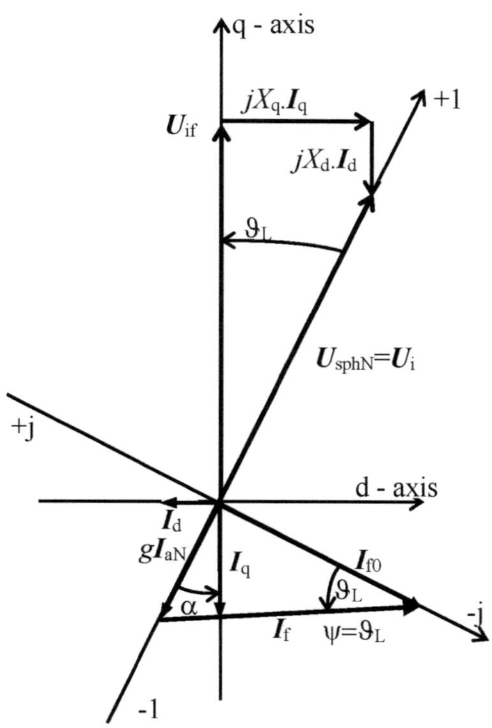

Figure 92.
Phasor diagram of synchronous machine for field current I_f and load angle ϑ_L calculation for the rated condition at $\cos\varphi = 1$.

An auxiliary angle α can be defined, which enables to calculate the load angle:

$$\frac{I_{f0}}{\sin\alpha} = \frac{I_f}{\sin 90^o} \Rightarrow \alpha = 48.1^\circ$$

Then the load angle in mechanical degrees is:

$$\vartheta_{\mathrm{Lmec}} = \frac{180 - 90 - 48.1}{p} = \frac{41.9}{2} = 20.95^\circ.$$

This is the value by which the rotor must be moved in the FEMM program for the rated condition. All needed values must be introduced to the program FEMM, and by the procedure described in Chapter 3.5.3, the value of electromagnetic torque is obtained. The calculated value of the electromagnetic torque 47.65 Nm and distribution of the magnetic flux lines are in **Figure 93**. The rated torque on the shaft calculated above is 47.74 Nm. It is seen again as a good coincidence between both values.

3.6.2 Reluctance synchronous machine

Calculation of the synchronous inductance in d- and q-axis and calculation of the air gap electromagnetic torque are shown. Analysis is made by means of the FEMM program on a real reluctance synchronous machine, the nameplate and basic parameters of which are in **Table 11**. The stator is identical with that of asynchronous motor 4AP90L, which was analyzed in Section 3.5. Therefore, the geometrical dimensions are taken from **Figure 67**. The rotor has also identical geometrical dimensions (**Figure 68**) but with such difference that 12 rotor teeth together, 3 for each pole, are cut off, to create salient poles (see **Figure 94**).

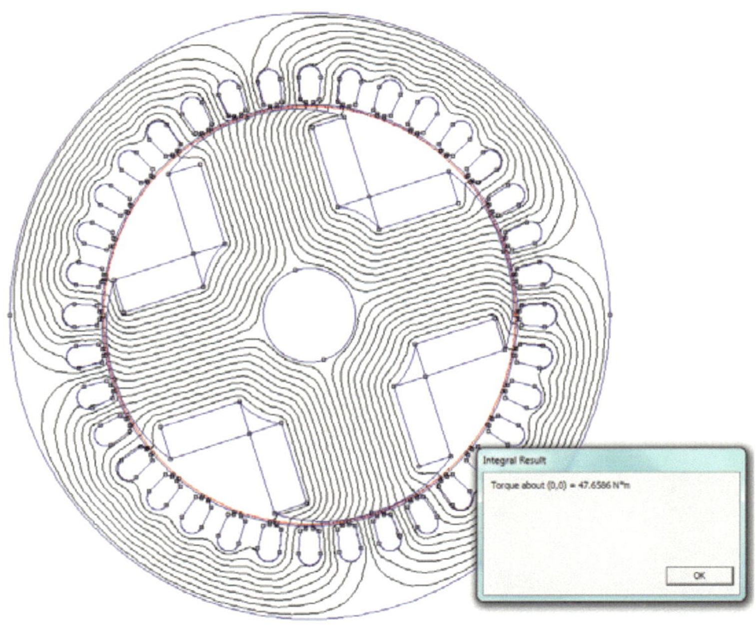

Figure 93.
The calculated torque and distribution of the magnetic flux lines in analyzed synchronous machine in rated condition, at cosφ = 1.

Rated stator voltage U_{1N}	400 V
Stator winding connection	Y
Rated power P_N	550 W
Rated frequency f	50 Hz
Rated speed n	1500 min^{-1}
Phase number m	3
Number of pole pairs p	2
Rated torque T_N	3.5 Nm
Rated load angle ϑ_L	7.25°
Number of turns of single-phase N_s	282
Number of slots per pole per phase q	3
Winding factor k_w	0.959
Active length of the rotor l_{Fe}	98 mm
Number of conductors in the slot z_Q	47
Magnetizing current (no load current) I_μ	3.22 A
Rated stator current I_{sN}	3.4 A
Leakage stator reactance (measured) $X_{\sigma s}$	5.4 Ω
Magnetizing d-axis reactance (measured) $X_{\mu d}$	69.44 Ω
Magnetizing q-axis reactance (measured) $X_{\mu q}$	33.2 Ω

Table 11.
Nameplate and parameters of the analyzed reluctance synchronous machine.

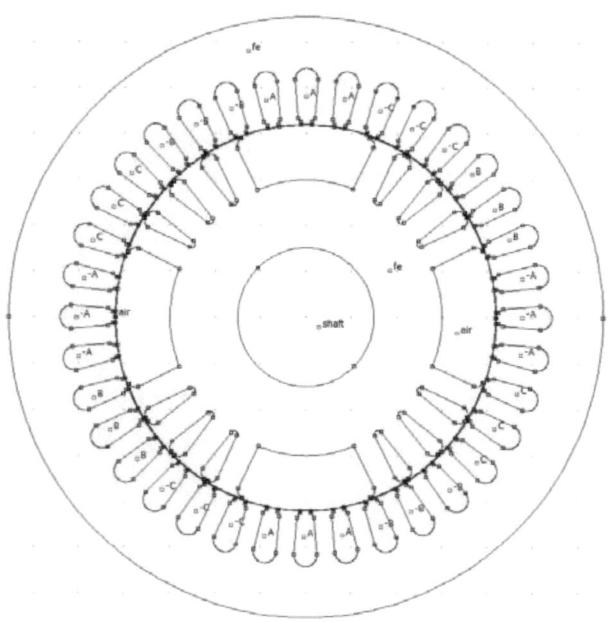

Figure 94.
Cross-section area of the analyzed reluctance synchronous machine (see also ***Figures 67*** *and* ***68****, where are geometrical dimensions of the stator and rotor).*

3.6.2.1 Calculation of the synchronous reactances X_d and X_q

Calculation of synchronous reactances is identical with that in synchronous machine with wound rotor, for d- and q-axis using Eqs. (494) and (495). The leakage stator reactance is taken from the asynchronous motor (Section 3.5); its measured value is in **Table 11** which is 5.4 Ω.

The machine is fed by three-phase stator current which corresponds to the magnetizing current (no load current) I_μ = 3.22 A. Rotor is set into position which corresponds to the d-axis. Again magnetostatic analysis is chosen, in which only one time instant is taken into account. The blocks representing ferromagnetic circuit are the same as in the case of asynchronous machine in Chapter 3.5, in which it means the same *B-H* curve for the sheets, with the thickness 0.5 mm marked Ei70. The current density calculation is made for the instant at which phase A current is zero, and the currents in phases B and C have values sin 60° from the magnitude of the current I_μ. If the sequence of phases and slots are according to **Figure 77**, then:

Phase A: A+ = 0, A− = 0
Phase C: C− = $-J_{\mu\max}\sin 60°$, C+ = + $J_{\mu\max}\sin 60°$
Phase B: B+ = $-J_{\mu\max}\sin 60°$, B- = + $J_{\mu\max}\sin 60°$

where $J_{\mu\max}$ is magnitude of the current density, calculated as follows:

$$J_{\mu\max} = \frac{z_Q I_{\mu\max}}{S_d} = \frac{47\cdot\sqrt{2}\cdot 3.22}{49.6} = 4.315\ \mathrm{MA/m^2}$$

where z_Q is number of the conductors in the slot and S_d is a surface of the slot. For q = 3 the sequence of the slots and phases are in **Figure 94**. The calculated current densities are set into the slots. The boundary condition is set on the zero magnetic vector potential A = 0 on the surface of stator, as it was in Chapter 5. In **Figure 95a**, there is distribution of the magnetic flux lines, and in **Figure 95b** the waveform of the air gap magnetic flux density along the whole air gap circumference. The magnitude of the fundamental harmonic component of the air gap magnetic flux density in d-axis is after the calculation by means of the Fourier series $B_{\delta 1d\max}$ = 0.9 T, which corresponds with the induced voltage according to Eq. (492), U_i = 222.5 V:

$$U_{i1d} = \sqrt{2}\pi f \frac{2}{\pi} B_{\delta 1\max}\frac{\pi D}{2p} l_{Fe} N_s k_w = \sqrt{2}\pi\frac{2}{\pi}0.9\frac{\pi.0.084}{4}0.098\cdot 282\cdot 0.959 = 222.5\ \mathrm{V}$$

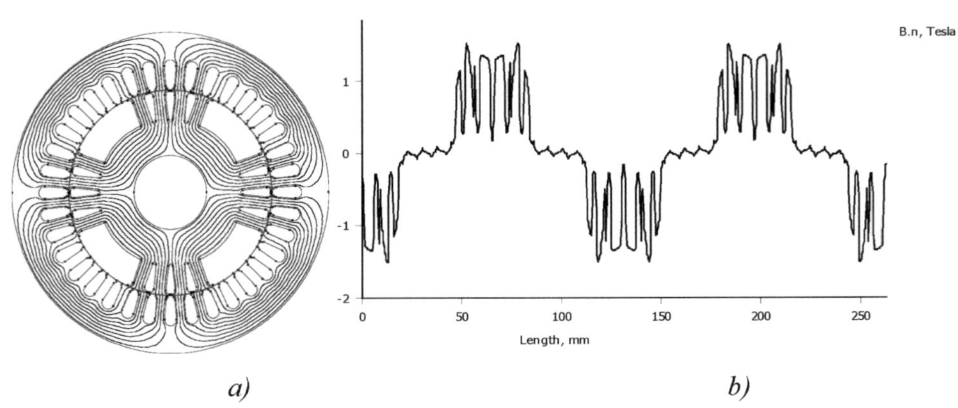

a) *b)*

Figure 95.
Outputs obtained from postprocessor for the d-axis, (a) distribution of the magnetic flux lines if the machine is fed by the current I_μ and (b) waveform of the magnetic flux density along the whole air gap circumference.

Calculation of the synchronous reactance in d-axis, according to Eqs. (497) and (494), is as follows:

$$X_{\mu d} = \frac{U_{i1d}}{I_{\mu d}} = \frac{222.5}{3.22} = 69.1\ \Omega$$

$$X_d = X_{\mu d} + X_{\sigma s} = 69.1 + 5.4 = 74.5\ \Omega.$$

The measured value is X_d = 74.84 Ω, which is in very good agreement with FEM analysis.

If synchronous reactance in q-axis is calculated, either the rotor or $B_{\delta 1max}$ is moved to the q-axis. Here it is done by moving the rotor. Unaligned axis of the rotor, i.e., q-axis (the axis between the poles), is put under the $B_{\delta 1max}$. The current can be the same I_μ = 3.22 A as for the d-axis, because in this position the big air gap dominates and there is no saturation of the ferromagnetic circuit. It means that the proportion between the induced voltage and magnetizing current is constant (Eq. (498)). A distribution of the magnetic flux lines in the air gap for the q-axis is in **Figure 96**. The magnitude of the fundamental component of the air gap magnetic flux density in the q-axis, obtained based on the Fourier series, is $B_{\delta 1qmax}$ = 0.424 T. The induced voltage in q-axis according to Eq. (492) is as follows:

$$U_{i1q} = \sqrt{2}\pi f \frac{2}{\pi} B_{\delta 1\,max} \frac{\pi D}{2p} l_{Fe} N_s k_w = \sqrt{2}\pi 50 \frac{2}{\pi} 0.424 \frac{\pi 0.084}{4} 0.098 \cdot 282 \cdot 0.959 = 104.84\ \text{V}$$

The magnetizing reactance and synchronous reactance in q-axis calculated according to Eqs. (498) and (495) are as follows:

$$X_{\mu q} = \frac{U_{i1q}}{I_{\mu q}} = \frac{104.84}{3.22} = 32.56\ \Omega$$

$$X_q = X_{\mu q} + X_{\sigma s} = 32.56 + 5.4 = 37.96\ \Omega$$

The q-axis synchronous reactance measured by means of the slip method is X_q = 38.6 Ω. It is seen in very good agreement between the measured and FEMM calculated values. Therefore, if there are accurate geometrical dimensions and material parameters, the FEM is a very good instrument for the investigation of the electrical machine parameters and properties.

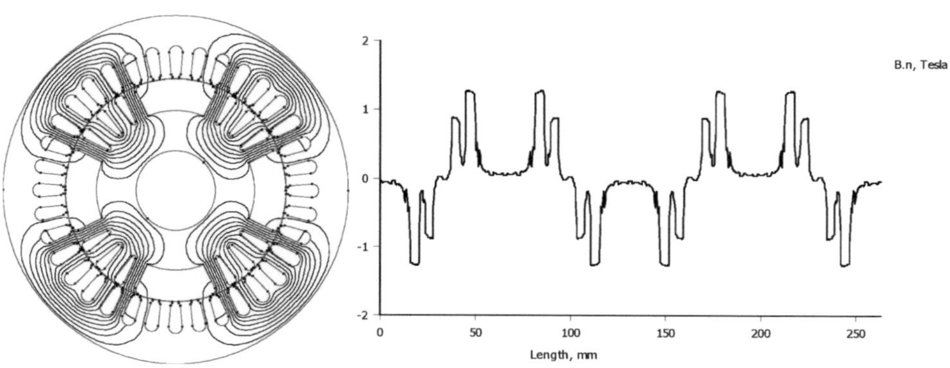

a) *b)*

Figure 96.
Outputs obtained from postprocessor for the q-axis, (a) distribution of the magnetic flux lines if the machine is fed by the current I_μ and (b) waveform of the magnetic flux density along the whole air gap circumference.

3.6.2.2 Calculation of the air gap electromagnetic torque

The air gap electromagnetic torque is calculated for the rated condition. Based on the nameplate values, the rated torque on the shaft can be calculated as:

$$T_N = \frac{P_N}{\Omega_N} = \frac{550}{\frac{2\pi n}{60}} = \frac{550}{\frac{2\pi 1500}{60}} = 3.5\ \text{Nm} \qquad (500)$$

In the stator slots, there is the rated current I_N = 3.4 A. The current density is calculated by means of the current magnitude (see expression (493)):

$$J_{Nmax} = \frac{z_Q I_{Nmax}}{S_d} = \frac{47 \cdot \sqrt{2} \cdot 3.4}{49.6} = 4.55\ \text{MA/m}^2$$

Again the magnetostatic case is employed; therefore, one time instant of the stator current is chosen, the best that at which phase A current is zero. Then the current densities in other phases are as follows:

$$\begin{aligned} J_{NB} &= -J_{Nmax} \sin 60^o = -4.55 \ \sin 60^o = -3.94\ \text{MA/m}^2 \\ -J_{NB} &= 3.94\ \text{MA/m}^2 \\ J_{NC} &= 3.94\ \text{MA/m}^2 \\ -J_{NC} &= -3.94\ \text{MA/m}^2 \\ -J_{NA} &= J_{NA} = 0 \end{aligned}$$

The value of the load angle is taken from the measurement and it is 7.25°. The rotor is moved by this value. It is possible to move the rotor gradually by other values of the load angle and to make the whole relation of the torque vs. load angle.

If all calculated needed parameters are entered to the FEMM program, the value of the electromagnetic torque is obtained. The procedure is similar as in Chapter 6.1.4. Distribution of the magnetic flux lines for this condition is in **Figure 97**. The calculated value of the air gap electromagnetic torque is 3.77 Nm. Rated torque

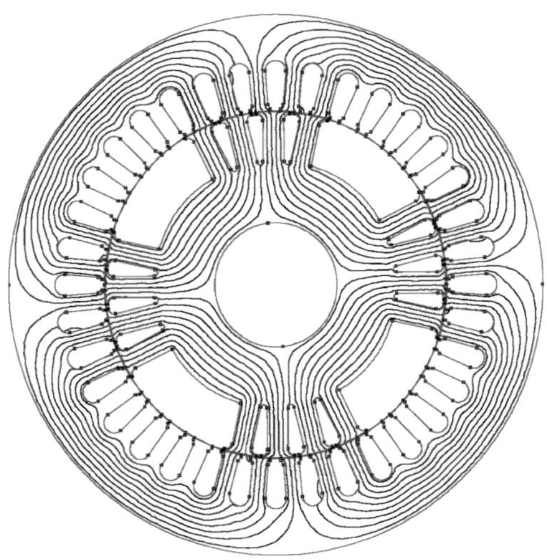

Figure 97.
Distribution of the magnetic flux lines in the analyzed reluctance synchronous machine for the rated load condition.

on the shaft, calculated above, is 3.5 Nm. Also here is in very good agreement between both values. By the way, the electromagnetic torque should be always higher than that on the shaft, because a part of the electromagnetic torque covers also no load losses.

3.6.3 Permanent magnet synchronous machines

Calculation of the induced voltage in no load condition, synchronous inductances in d-axis and q-axis, air gap electromagnetic torque, and the so-called cogging torque (defined in Section 6.3.4) is made in this chapter for the permanent magnet synchronous machines (PMSM), by means of FEM. A real PMSM, the cross-section area of which is in **Figure 98**, is analyzed. In **Figure 99** there is a stator slot of this machine with its geometrical dimensions. Double-layer stator winding is employed. The 12 NdFeB magnets, which are buried (embedded in the surface) around the whole rotor circumference, are applied. The nameplate and other parameters of the analyzed machine are in **Table 12**.

3.6.3.1 Induced voltage calculation in no load condition

The aim of the PMSM simulation is to calculate induced voltage for the defined speed, in this case for the rated speed 360 min^{-1}, which results in the rated frequency 36 Hz. The procedure is similar like in the case of synchronous machine with wound rotor (Chapter 6.1), but here the permanent magnets are employed. These are defined in *Properties–Materials* by means of magnet coercitive force H_C. The dialog window for permanent magnet setting is in **Figure 100**.

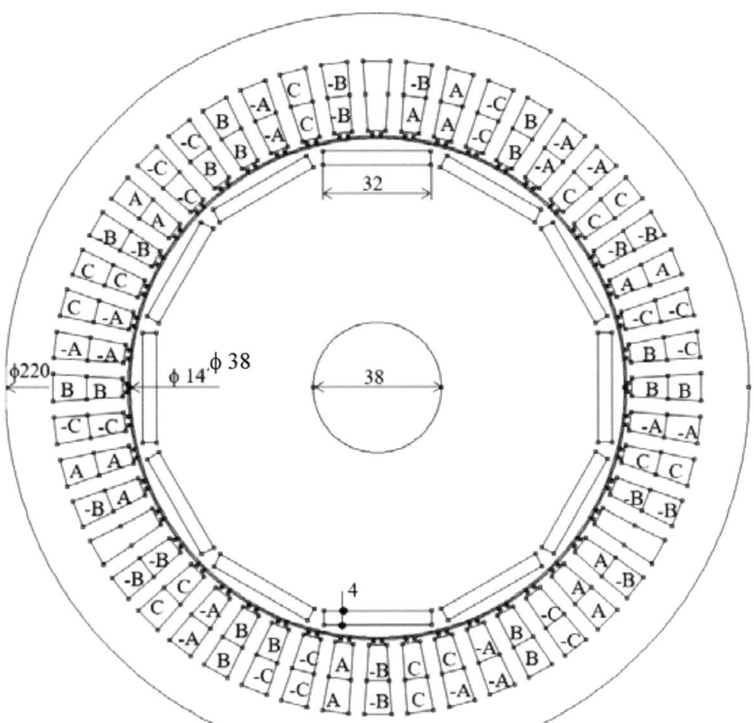

Figure 98.
Cross-section area of the analyzed PMSM.

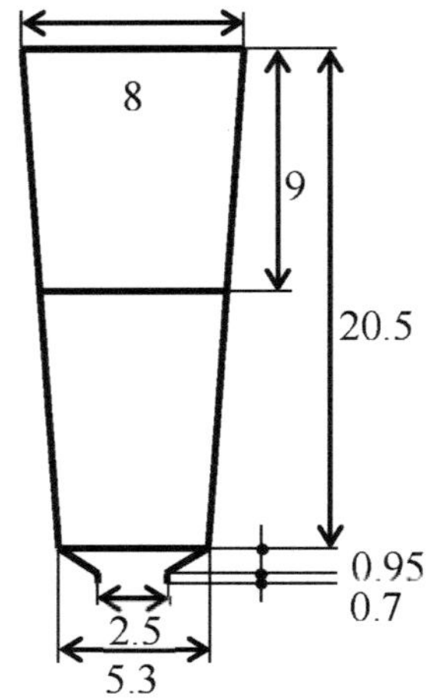

Figure 99.
Stator slot of the analyzed PMSM and its geometrical dimensions.

Rated stator voltage U_{phN}	133 V
Stator winding connection	Y
Rated frequency f	36 Hz
Rated speed n	360 min^{-1}
Rated power P_N	3000 W
Phase number m	3
Number of pole pairs p	6
Rated torque T_N	80 Nm
Rated load angle ϑ_L	14° mech.
Number of turns in single-phase N_s	420
Number of slots per phase per pole q	1.25
Winding factor k_w	0.936
Active length of the rotor l_{Fe}	140 mm
Number of conductors in the slot z_Q	56
Number of stator slots Q_s	48
Rated stator current I_{sN}	8.3 A
Number of stator slots with embedded conductors Q	45
Remanent magnetic flux density of NdFeB magnets B_r	1.11 T
Coercitive force H_c	850 kA/m
Air gap δ	0.8 mm
Leakage reactance (measured) $X_{\sigma s}$	11.3 Ω

Table 12.
Nameplate and parameters of the analyzed PMSM.

The boundary conditions are set in a similar way like in asynchronous or synchronous machine with wound rotor. It is supposed that zero magnetic vector potential is $\mathbf{A}$ = 0 on the stator surface. Then it is possible to launch mesh creation and calculation. Close to the air gap, the mesh can be refined to get the results more

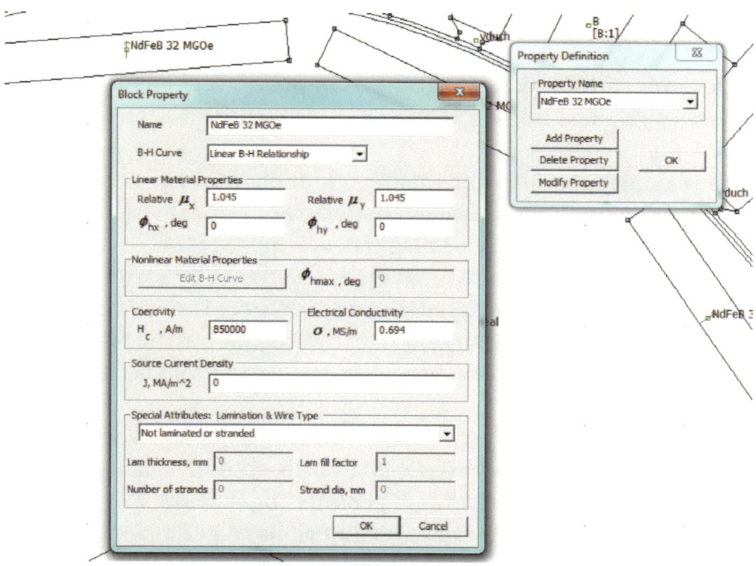

Figure 100.
Settings of the permanent magnet parameters in FEMM program.

accurate. After the calculation is finished, distribution of the magnetic flux lines is obtained (**Figure 101a**). In *postprocessor*, mark an auxiliary circle in the middle of the air gap, and ask to describe waveform of air gap magnetic flux density, around the whole circumference (**Figure 101b**). Fourier series enables to calculate all harmonic components, but for calculation of the induced voltage, the fundamental component of the air gap magnetic flux density, $B_{\delta 1max}$ = 0.66 T, is used.

Now it is possible to calculate a phase value fundamental component of the induced voltage:

$$U_{i1} = \sqrt{2}\pi f \frac{2}{\pi} B_{\delta 1\,\max} \frac{\pi D}{2p} l_{Fe} N_s k_w = \sqrt{2}\pi 36 \frac{2}{\pi} 0.66 \frac{\pi 0.1476}{6} 0.14 \cdot 420 \cdot 0.936 = 143\ \mathrm{V}$$

The calculated value is in appropriate agreement with the nameplate value, which is 133 V. Inaccuracy can be caused by *B-H* curve which is unknown and was taken from the program library.

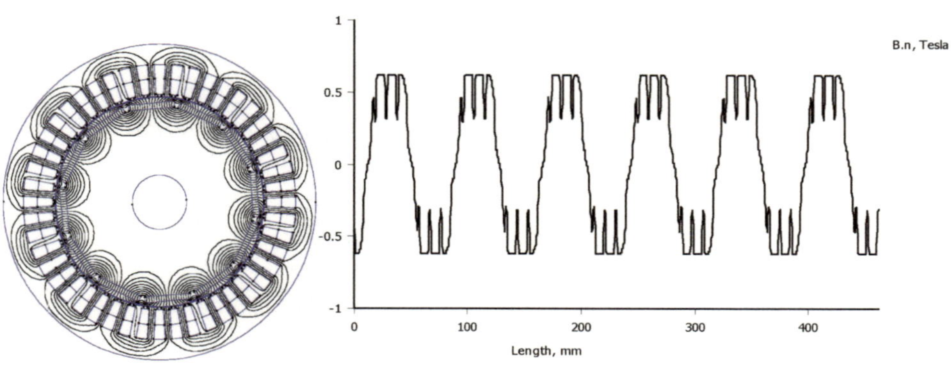

a) *b)*

Figure 101.
Outputs from the postprocessor, (a) distribution of the magnetic flux lines and (b) waveform of the magnetic flux density around the whole air gap circumference.

3.6.3.2 Synchronous reactances X_d and X_q calculation

A procedure for calculation is the same as in previous chapters, based on Eqs. (494) and (495). The leakage inductance is taken from a measurement and is $L_{\sigma s}$ = 50 mH.

During the calculation of the magnetizing inductances, the permanent magnets are replaced by nonmagnetic materials, the relative permeability of which corresponds with relative permeability of magnets μ_r = 1.045, [11].

With regard to the fact that the magnets are replaced by material close to vacuum, the calculation of the magnetizing inductances is recommended [11] for 10% of rated current to get unsaturated inductance and for the 80–100% of rated current to get saturated inductance. Here the calculation for the rated current is shown.

The machine is fed by three-phase rated current I_N = 8.3 A. The rotor is set to the d-axis position. Again the magnetostatic analysis is chosen because only one time instant is analyzed. The case of double-layer winding is more suitable to feed the slots by "amperturns." The amperturns setting is made in *Properties/Circuits*, where turns connection is set as series and time instant of the current is chosen at zero current in phase A. The currents in phases B and C are as follows:

Phase A: A+ = 0, A− = 0
Phase C: C− = $-(z_Q/2)\,I_{Nmax}\sin 60^\circ$, C+ = $+(z_Q/2)\,I_{Nmax}\sin 60^\circ$
Phase B: B+ = $-(z_Q/2)\,I_{Nmax}\sin 60^\circ$, B− = $+(z_Q/2)\,I_{Nmax}\sin 60^\circ$

where $z_Q/2$ is the number of the conductors in one layer of the slot. An example of the calculation is as follows:

$$\left(\frac{z_Q}{2}\right) I_{\mathrm{Nmax}} \sin 60^\circ = \frac{56}{2}\sqrt{28.3}\sin 60^\circ = 284.6 \text{ Amperturns.} \tag{501}$$

A sequence of the slots and phases is shown in **Figure 98**. The calculated values of amperturns are set in those slots. The boundary condition is set to zero magnetic vector potential **A** = 0 on the surface of the stator. In **Figure 102a** there is distribution of the magnetic flux lines, and in **Figure 102b** there is a waveform of the air gap magnetic flux density around the whole circumference.

The magnitude of the air gap magnetic flux density is calculated based on the Fourier series, and in d-axis, it is $B_{\delta 1maxd}$ = 0.64 T. According to Eq. (492), the induced voltage in d-axis is:

$$U_{i1d} = \sqrt{2}\pi f \frac{2}{\pi} B_{\delta 1\,\mathrm{max}\,d} \frac{\pi D}{2p} l_{\mathrm{Fe}} N_s k_{\mathrm{w}} = \sqrt{2}\pi 36 \frac{2}{\pi} 0.64 \frac{\pi .0.1476}{12} 0.14 \cdot 420 \cdot 0.936 = 138.6 \text{ V}$$

Then the synchronous reactance in d-axis can be calculated by means of Eq. (497) and (494):

$$X_{\mu\mathrm{d}} = \frac{U_{\mathrm{i1d}}}{I_{\mu\mathrm{d}}} = \frac{138.6}{8.3} = 16.7\ \Omega$$

$$X_{\mathrm{d}} = X_{\mu\mathrm{d}} + X_{\sigma\mathrm{s}} = 16.7 + 11.3 = 28\ \Omega$$

The calculation of the synchronous reactance in q-axis is made by setting the rotor into the q-axis. It means that the unaligned space between the poles (q-axis) is under the magnetic flux density magnitude $B_{\delta 1max}$. The calculation is made for the rated current. A distribution of the magnetic flux lines in q-axis is shown in

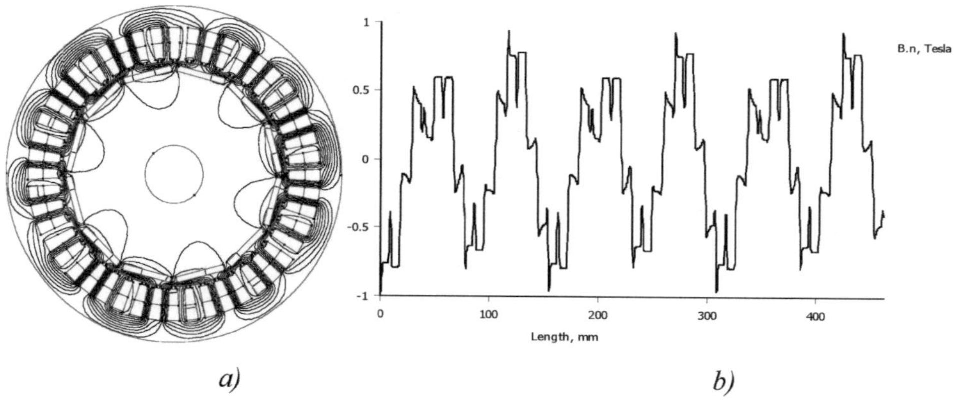

Figure 102.
(a) Outputs obtained from the postprocessor for d-axis if it is fed by the rated current and (b) waveform of the magnetic flux density around the whole air gap circumference.

Figure 103a. Based on the Fourier series, the magnitude of the air gap magnetic flux density in q-axis is $B_{\delta 1 \max q}$ = 0.732 T. According to Eq. (492), the induced voltage in single phase in q-axis can be calculated:

$$U_{\mathrm{i1q}} = \sqrt{2}\pi f \frac{2}{\pi} B_{\delta 1 \max q} \frac{\pi D}{2p} l_{\mathrm{Fe}} N_{\mathrm{s}} k_{\mathrm{w}} = \sqrt{2}\pi 36 \frac{2}{\pi} 0.732 \frac{\pi.0.1476}{12} 0.14 \cdot 420 \cdot 0.936 = 158.5\ \mathrm{V}$$

The magnetizing and synchronous reactance in q-axis is calculated by means of Eqs. (498) and (495):

$$X_{\mu \mathrm{q}} = \frac{U_{\mathrm{i1q}}}{I_{\mu \mathrm{q}}} = \frac{158.5}{8.3} = 19.09\ \Omega$$

$$X_{\mathrm{q}} = X_{\mu \mathrm{q}} + X_{\sigma \mathrm{s}} = 19.09 + 11.3 = 30.39\ \Omega.$$

As it is known from the theory of PMSM with the permanent magnets buried in the rotor, the synchronous reactance in q-axis is higher than that in the d-axis, $X_q > X_d$, which is in coincidence with the obtained results.

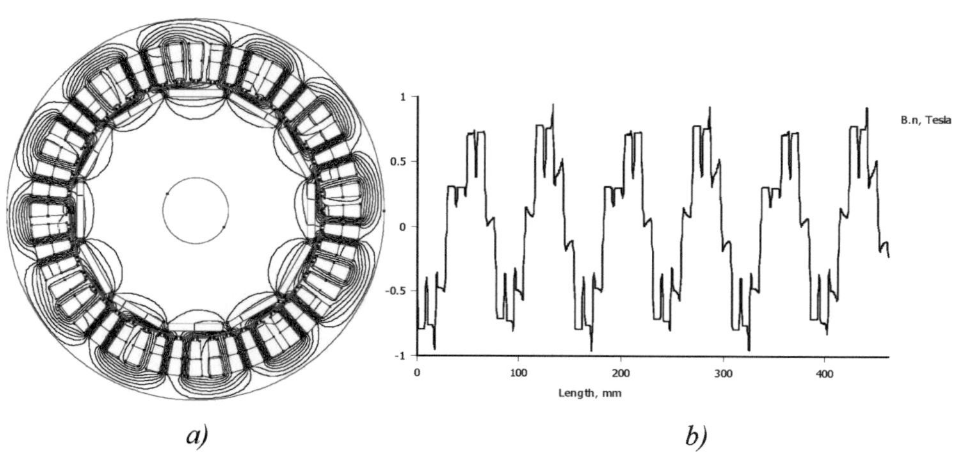

Figure 103.
(a) Outputs obtained from the postprocessor for q-axis if it is fed by the rated current and (b) waveform of the magnetic flux density around the whole air gap circumference.

3.6.3.3 Calculation of the electromagnetic torque in the air gap

The calculation of the air gap electromagnetic torque of the PMSM is made for the rated condition based on the nameplate values.

The rated current in the stator slots is I_N = 8.3 A. A procedure of the current settings in the slots is the same as in Section 6.3.2. The value of the load angle taken from measurement is 14° mech. The rotor must be moved by this value.

If all needed values and parameters are entered to the FEMM program, following the procedure in Section 6.1.4, the value of electromagnetic torque is obtained. A distribution of the magnetic flux lines for this condition is in **Figure 104**. The calculated value of the air gap electromagnetic torque is 84 Nm. The rated torque on the shaft in **Table 12** is 80 Nm. It is seen in very good agreement between both values, taking into account that if the electromagnetic torque in the air gap covers also the torque loss; therefore, the torque on the shaft is lower.

3.6.3.4 Calculation of the cogging torque in the air gap

Cogging torque T_{cogg} is a torque developed by interaction of permanent magnets and stator slotting [4]. It is also known as the torque without current. It means that it is torque developed only by permanent magnets if no current flows in the stator winding. It depends on rotor position and periodically is changing depending on the pole number $2p$ and stator slot number. In the motoring operation, it is unwanted phenomenon. It is felt mainly at the low speed by jerky movement of the rotor, i.e., by the ripple torque. At high speed, its influence on the speed is filtered by the rotor moment of inertia.

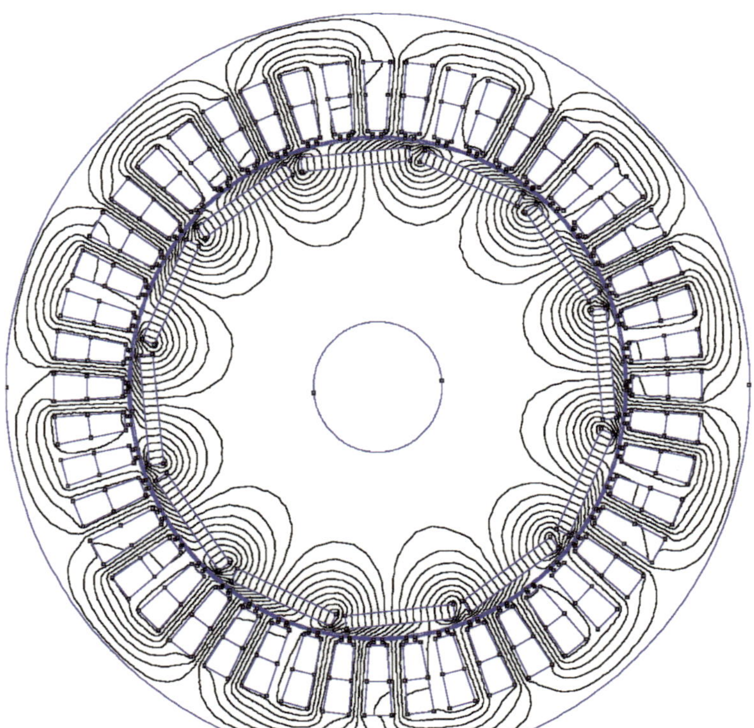

Figure 104.
Distribution of the magnetic flux in analyzed PMSM for the rated condition.

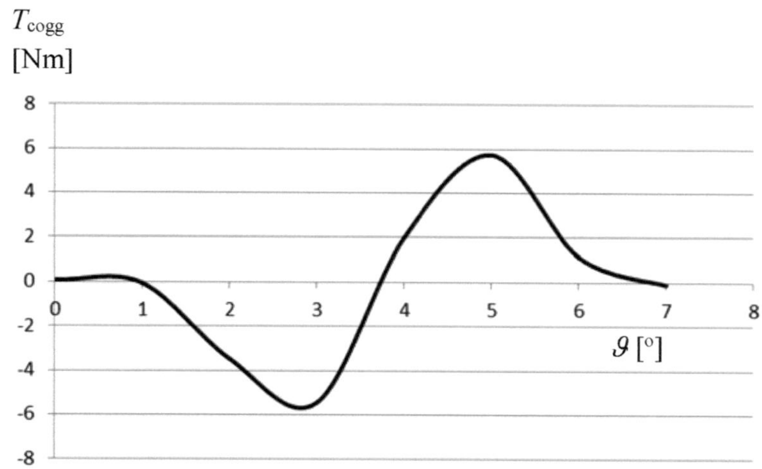

Figure 105.
Waveform of the cogging torque of the PMSM vs. rotor position.

Its calculation is made by such a way that the stator winding is current-free. The rotor is moved gradually, and based on the Maxwell stress tensor, the air gap torque is investigated. Then it is described in dependence on the rotor position. The waveform of here investigated cogging torque is shown in **Figure 105**. It is seen that its maximum is 6 Nm, which is 7.5% from the rated torque.

3.7 Analysis of the switched reluctance motor parameters

Switched reluctance motor (SRM) is from the point of view of parameters determination that is totally different in comparison with asynchronous and synchronous machines. Here an analysis of SRM parameters by means of the FEM that is carried out. The cross-section area of the investigated SRM with its geometrical dimensions is in **Figure 106**. The nameplate and other known parameters are in **Table 13**.

The aim of this chapter is to show how the static parameters of the SRM can be calculated, such as phase inductance, linkage magnetic flux, coenergy, and electromagnetic torque. All mentioned parameters should be calculated for all important rotor positions and all current values which can occur during the operation. The main difference in comparison with other kinds of the electrical machine is that the stator winding is not distributed but it is wound concentrically on the stator teeth and the rotor is without any winding, which makes investigation by means of FEM easier. The electromagnetic circuit of the SRM is possible to investigate as a coil with ferromagnetic core and with the air gap (see Section 3). The FEM program enables the calculation of inductances, magnetic fluxes, coenergies, and torques in such simple circuit directly without any further recalculation.

The basic input operations for FEMM program calculation are as follows:

- Drawing of the cross-section area based on its geometrical dimensions
- Setting of material constants and *B-H* curves
- Setting of the fed phase current density (calculated based on the turns number and phase current)
- Setting of the boundary conditions.

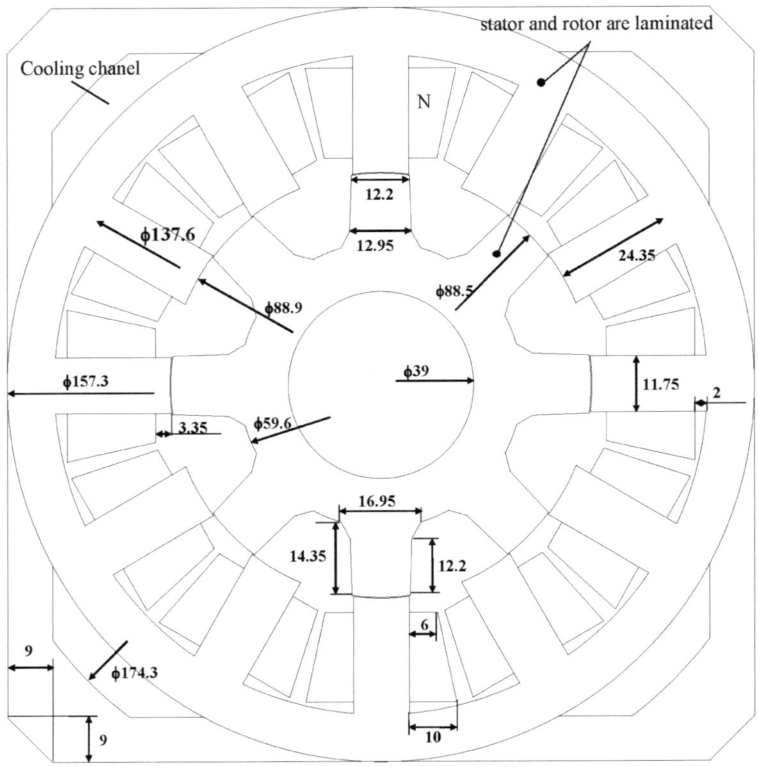

Figure 106.
Cross-section area of the investigated SRM with its geometrical dimensions.

Voltage of the DC link U_{DC}	540 V
Rated speed n	3000 min^{-1}
Rated power P_N	3700 W
Phase number m	3
Stator teeth number N_s	12
Rotor teeth number N_r	8
Type	12/8
Turns number in a single-phase N_s	49
Number of the coils connected in series creating one phase	4
Rated torque T_N	11.8 Nm
Active length of the rotor l_{Fe}	114.5 mm
Resistance of single-phase R_{ph}	0.61 Ω
Air gap length δ	0.2 mm

Table 13.
Nameplate and parameters of the investigated SRM.

The results accuracy depends on the number of the finite elements used in the cross-section area and on the accuracy of the input parameters. Because in this type of machine, a lot of calculations is required, the number of the finite elements must ensure a sufficient accuracy for relatively short time. The further increasing of the number of elements does not increase the accuracy considerably, but it increases only the calculation time. The calculations are static; it means that they are made for one rotor position and one concrete constant current.

It can be done in two ways. The first one, which can take a lot of time, because it is necessary to set gradually the values of the current (between 1 and 30 A) for all rotor positions. The rotor positions are changed gradually by 1° mech, from aligned to unaligned positions, which is, in this case, 22.5°. Therefore, it would be needed to carry out about 690 calculations.

The second approach is based on the possibility to create a parametrical model of the cross-section area, without very complicated drawing of the newer and newer cross-section areas. The FEMM program enables to use LUA script program, in which a computing loop can be created, and to define the whole calculation in the loop. It means that the rotor position is gradually changed at the constant current and then the phase current is changed. A printout of the program for the parametrical SRM model in LUA script is in Appendix D. In **Figure 107**, there is a distribution of the magnetic flux lines for two limit positions for phase A: **Figure 107a**, aligned position, and **Figure 107b**—unaligned position. As it is seen in the figures, the lines enter also to the shaft. It is given by the fact that the *B-H* curves of the ferromagnetic sheets and shaft are similar.

3.7.1 Calculation of the linkage magnetic flux

The linkage magnetic flux in FEMM is calculated based on the equation:

$$\phi = \oint_l \boldsymbol{A} \mathrm{d}l, \tag{502}$$

where $\boldsymbol{A}$ is magnetic vector potential and l is a circumference of the surface, where the vector A is calculated. The boundary condition is taken as $\mathbf{A} = 0$ on the border of the stator. The linkage magnetic flux ψ is a product of the magnetic flux ϕ and the number of turns N_s, by which it is linked.

The calculated curves $\psi = f(i, \vartheta)$ and $\psi = f(\vartheta, i)$ are in **Figures 108** and **109**. In **Figure 110** there are curves obtained from FEMM program and from measurements. The obtained results confirm that the differences in aligned position are not higher than 3%, mainly in the saturated region, which is in very good agreement. It depends on the accuracy of the *B-H* curve. In the unaligned position, this difference is higher, about 12%, which is caused by the fact that in this position, a leakage of

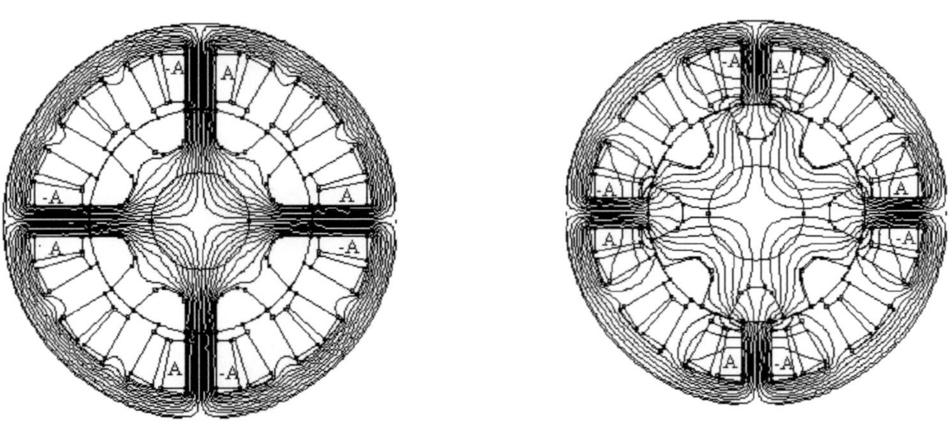

a) b)

Figure 107.
Distribution of the magnetic flux lines for (a) aligned position and (b) unaligned position.

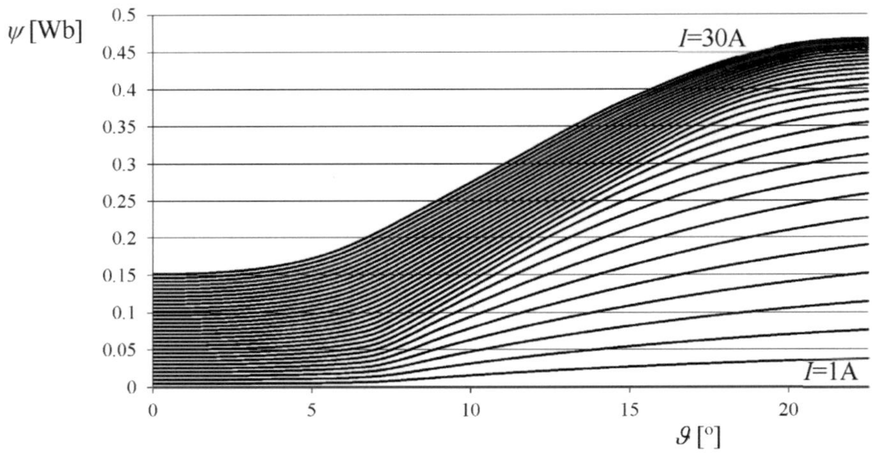

Figure 108.
Linkage magnetic flux vs. rotor position and current ψ = f(ϑ, I), current is changed from 1 to 30 A with the step 1 A.

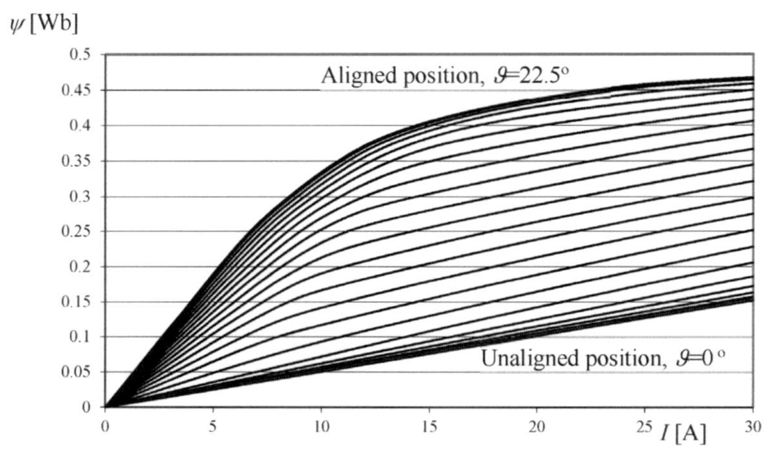

Figure 109.
Linkage magnetic flux vs. current and rotor position ψ = f(I, ϑ).

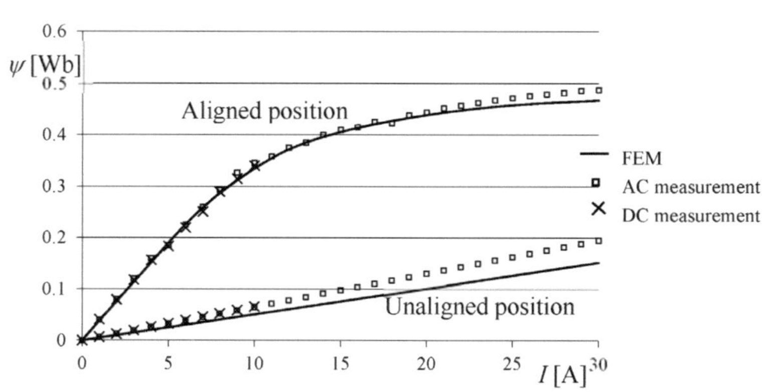

Figure 110.
Linkage magnetic flux vs. current and rotor positions ψ = f(i, ϑ) for aligned and unaligned positions obtained by measurements and by means of the FEM.

the end winding is more significant than in the aligned position, which is not taken into account in the 2D analysis. In **Figure 110** it is seen that the FEM program is a reliable method to determine these curves.

3.7.2 Inductance of single-phase calculation

The inductance can be calculated by two expressions. At first it is the expression for the linkage magnetic flux:

$$\psi = L_{\mathrm{ph}} i \Rightarrow L_{\mathrm{ph}} = \psi / i,$$

which is valid in linear and nonlinear case and which is used below. The second expression results from the energy of electromagnetic field in linear case, which gives:

$$W = \frac{1}{2} L_{\mathrm{ph}} i^2 \Rightarrow L_{\mathrm{ph}} = \frac{2W}{i^2}. \qquad (503)$$

In linear region, the inductances are equal in both ways of calculation. In nonlinear region, there are differences.

In **Figure 111** there are waveforms of the phase inductance $L_{\mathrm{ph}} = \mathrm{f}(\vartheta, I)$, only for single phase. Because of symmetry of electric and magnetic circuits, it is supposed that there are the same waveforms for other phases and there is no need to calculate them. The mutual inductance can be neglected, as it is mentioned by many authors. It was confirmed also by FEM calculation in this investigated case. It was found out that the mutual inductances are lower order position and can be neglected.

3.7.3 Coenergy calculation

In Section 3.2, it was derived that electromagnetic torque can be calculated based on coenergy. In the FEM program, the coenergy can be calculated by means of the expression:

$$W' = \int_{\Omega} \left(\int_0^H B \mathrm{d}H \right) \mathrm{d}S \qquad (504)$$

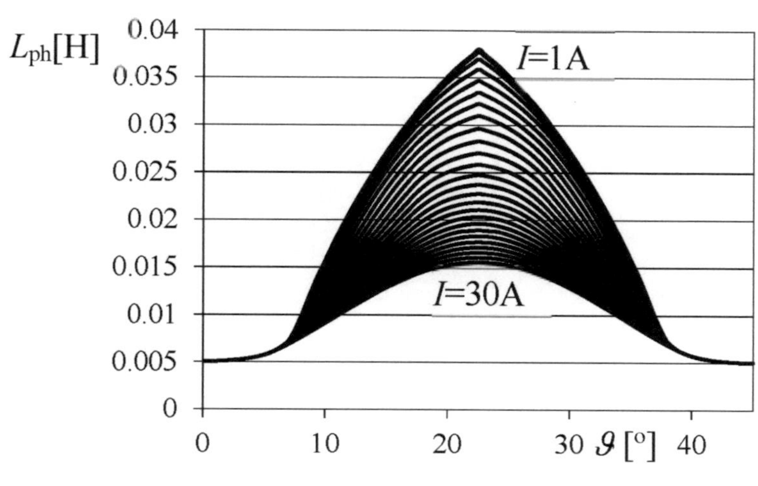

Figure 111.
Waveforms of the phase inductance L = f(ϑ, I), the current is changed from 1 A to 30 A by a step 1 A.

where S is a region on which the coenergy is integrated. In this case this region means cross-section area. Equation (504) can be modified as follows:

$$W' = \int_{\Omega} \left(\int_{0}^{i} \psi \mathrm{d}i \right) \mathrm{d}S. \qquad (505)$$

In **Figure 112** there are curves of coenergy vs. current and rotor positions $W' = \mathrm{f}(I, \vartheta)$, which can be obtained by a gradually changed rotor position and for various values of the phase current.

3.7.4 Static electromagnetic torque of the SRM

The static electromagnetic torque of the SRM is possible to calculate by means of the FEM, based on Eqs. (473) and (476).

The accuracy of FEM calculation depends on the value of rotor moving $\Delta\vartheta$. The lower the value of the $\Delta\vartheta$, the higher the precision of the results, but the calculation takes more time. The step of $\Delta\vartheta = 1°$ mech. was chosen for calculation here, on the interval between the aligned and unaligned position, which seemed to be appropriate. The calculated values $T = \mathrm{f}(\vartheta, I)$ are in **Figure 113** for the SRM 12/8.

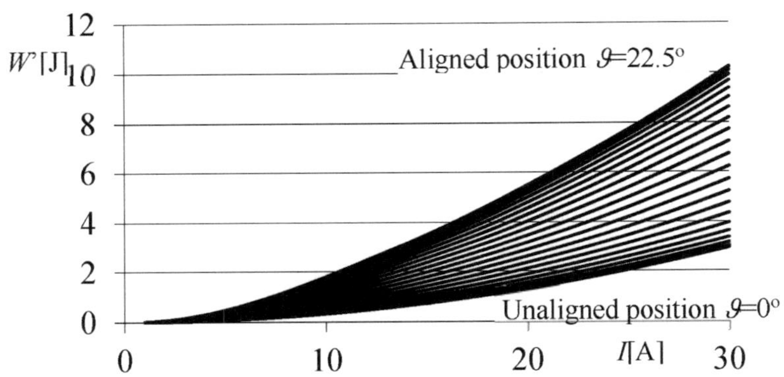

Figure 112.
Coenergy vs. current and rotor position W' = *f*(I, ϑ) *for the SRM 12/8.*

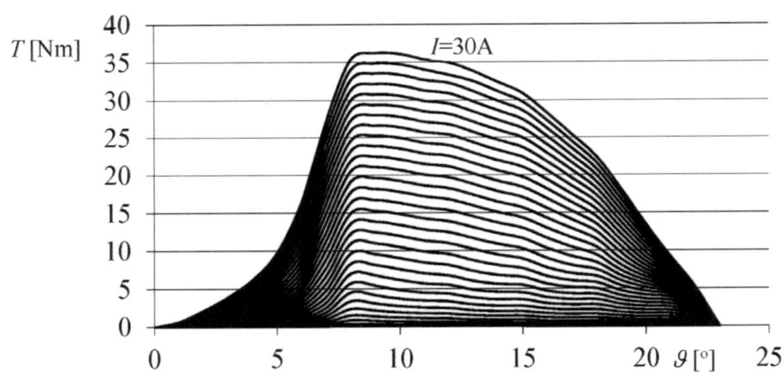

Figure 113.
Static torque curves vs. rotor position and current T = *f*(ϑ, I) *for the SRM 12/8, the current is changed from 1 A to 30 A with the step of 1 A.*

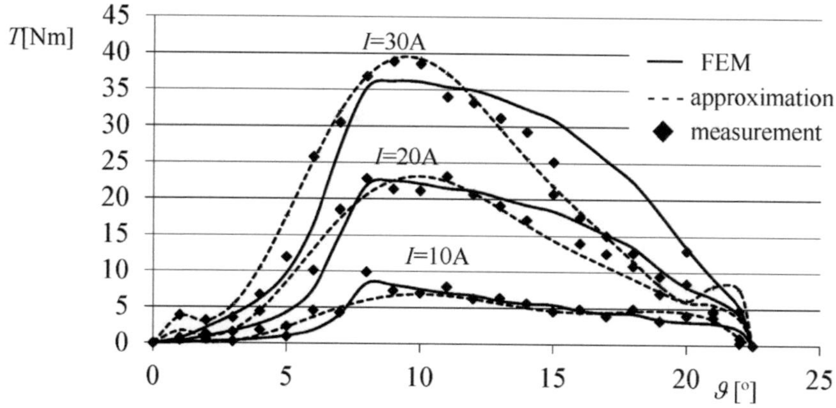

Figure 114.
Static torque curves vs. rotor position and currents T = *f*(ϑ, I) *for currents* I = 10, 20, 30 *A of the SRM 12/8.*

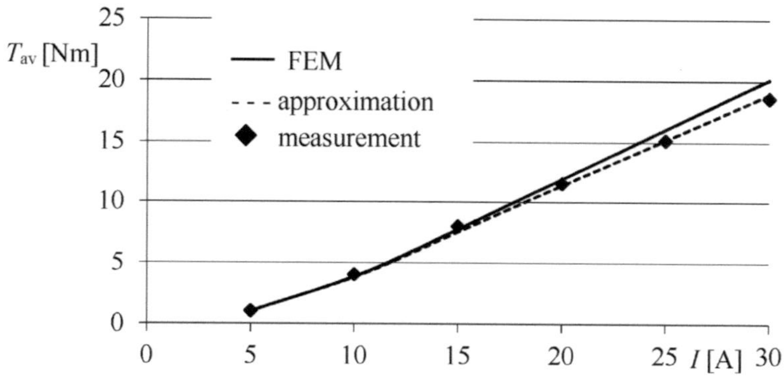

Figure 115.
Waveform of the average torque vs. current $T_{av} = f(I)$*, obtained from measured and FEM calculated values. The measured values are approximated.*

In **Figure 114** there are static torque curves vs. rotor position for the currents I = 10, 20, 30 A, due to better transparency, obtained by measurements and by FEM calculation. The measured values are approximated to see the waveforms.

These static torque curves are the basis for the calculation of the average torque T_{av} by means of the equation:

$$T_{av} = \left[\int_{\vartheta=0}^{\vartheta=22.5} T d\vartheta \right]_{i=const.} \tag{506}$$

This calculation results in the values obtained for measured values and FEM calculated values. Both are shown in **Figure 115** for the currents I = 5, 10, 15, 20, 25, 30 A.

In **Figure 115** it is seen that a difference between calculated and measured values exists. It is supposed that 3D FEM simulation, more accurate *B-H* curves, and measurements would decrease the differences.

Based on all obtained outputs, it can be proclaimed that FEM simulation is in very good instrument for electrical machine analysis and can be used in prefabricated period of their design to optimize their properties.

Symbols

A	magnetic vector potential
a	number of parallel paths in the armature
b	slot width
B	magnetic flux density
B_r	remanence flux density
C	DC machine factor
E	electric field strength
D	electric flux density, diameter
F	force
k	factor, coefficient
h	slot height
H	magnetic field strength
H_c	coercitivity of PM
i	current, instantaneous value
J	moment of inertia
J	current density
l	length
L	inductance
L_d	slot leakage inductance, synchronous inductance in d-axis
L_q	synchronous inductance in q-axis
L_z	tooth tip leakage inductance
L_{cc}	end winding leakage inductance
L_{sp}	pole leakage inductance
l_{Fe}	active length of the iron
T	torque
M	number of phases
N	rotation speed, ordinal of the harmonic, ordinal of the critical rotation speed, number of rotor bars
N	number of turns in a winding
P	number of pole pairs
p_e	electromagnetic power, instantaneous value
P	power
ΔP	losses
Q	electric charge
Q_R	number of rotor bars
Q	number of slots per pole and phase
R	resistance
R	unit vector
S	surface, apparent power
S	slip
T	time, temperature
t_e	electromagnetic torque, instantaneous value
T_t	periods, time constant
U	rms voltage, depiction of the phase U
V	volume, depiction of phase V
W	energy stored in magnetic field, depiction of the phase W
W'	coenergy
U	voltage, instantaneous value
$X_{\sigma s}$	leakage reactance of stator
$X_{\mu d}$	magnetizing reactance in d-axis

$X_{\mu q}$	magnetizing reactance in q-axis
X_d	synchronous reactance in d-axis
X_q	synchronous reactance in q-axis
Y	winding pitch, star connection
z_a	number of adjacent conductors in the slot
z_t	number of conductors on the top of each other in the slot
z_Q	number of the conductor in the slot
α	mutual rotation of currents in rotor bars
α_i	ratio of the magnetic flux densities B_{av}/B_{dmax}
γ	phase shift of the top and bottom layer
δ	air gap
μ	permeability
ϑ	angle
ϑ_r	rotor position
ϑ_L	static angle of load
λ	permeance factor
Λ	magnetic permeance
ξ	winding factor, referred height of the conductor
σ	specific conductivity
τ	time constant
τ_p	pole pitch
ν	ordinal of harmonic
ϕ	magnetic flux
ψ	magnetic flux linkage
ω	electric angular speed
ω_s	electric angular speed of stator field
Ω	mechanical angular speed

Subscripts

0	linked with no load condition
a, b, c	phases A, B, C
Al	Aluminum
AC	alternating current value
av	average value
cc	linked with the end winding
Cu	copper
cv	windings
D	d-axis of the rotor
d	axis d, direct axis; linked with the slot
DC	direct current value
e_m	electromagnetic
hn	driving torque
f	excitation winding
ph	phase value
Fe	linked with ferromagnetic
j	general expression
k	k-order
m	number of phases, mass
mag	magnetic
max	maximal
min	minimal
N	nominal, rated

Q	axis q of the rotor
q	axis q, quadrature axis, armature winding
p	primary
r	linked with the rotor
R	linked with the electric resistance
rms	the square root of the arithmetic mean of the squares of the values
rot	rotating
ripp	ripple
s	synchronous, linked with the stator, secondary
S	linked with the stator
sq	slot skewing
tr	transforming
vs	stator winding
z	linked with the tooth of the machine
L	load
α	α-*axes*
β	β-axes
δ	linked with the air gap
μ	magnetizing
ν	harmonic component order
σ	leakage

A. Appendix

A.1 Simulation model of separately excited DC motor in MATLAB-Simulink

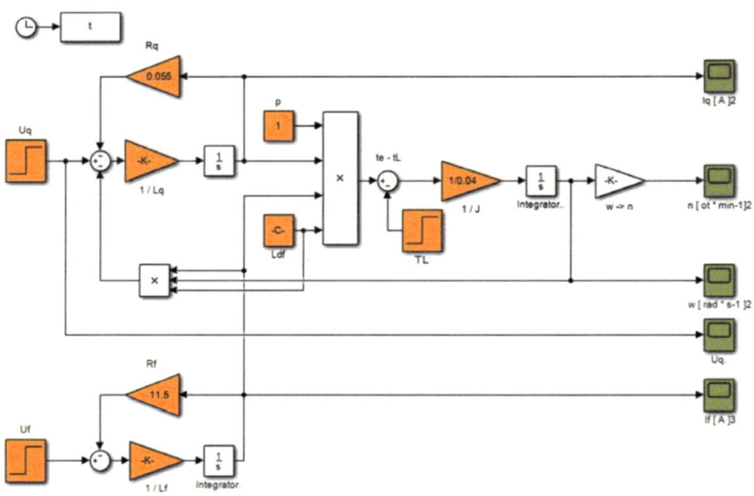

A.2 Simulation model of separately excited DC generator in MATLAB-Simulink

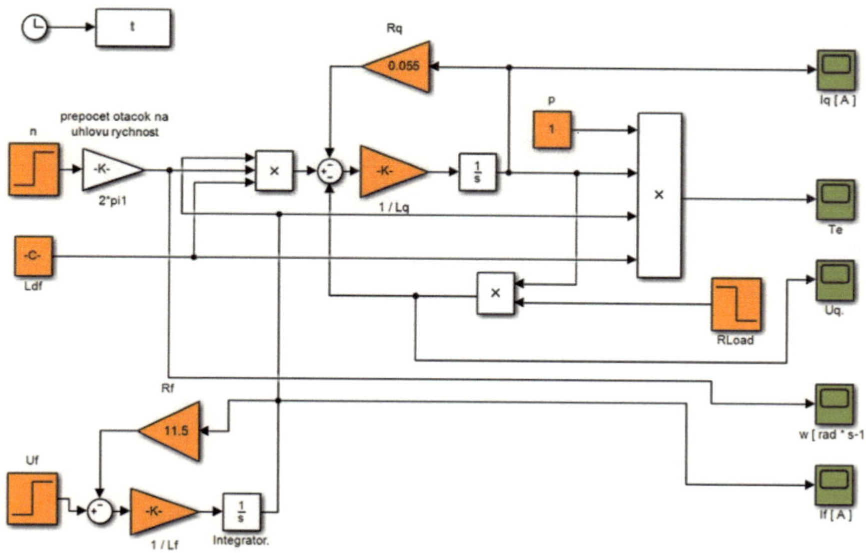

A.3 Simulation model of shunt wound DC motor in MATLAB-Simulink

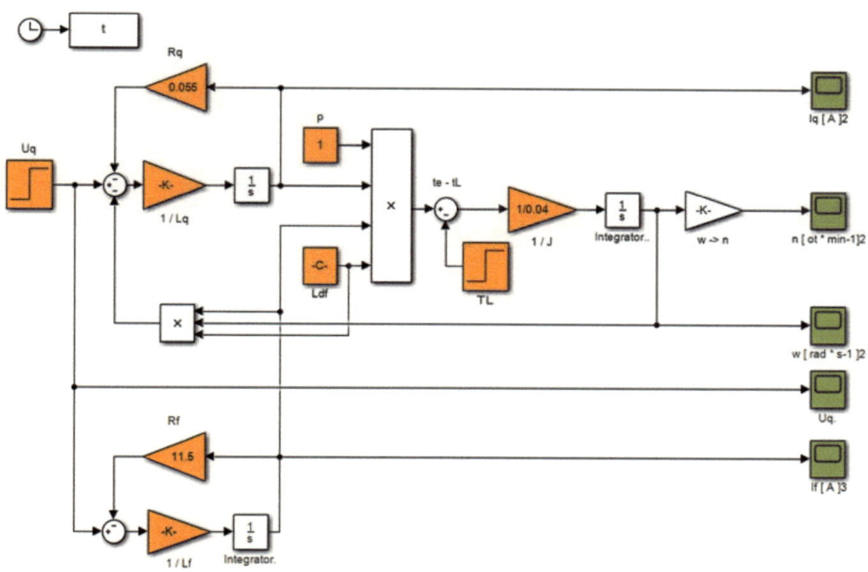

A.4 Simulation model of shunt wound DC generator in MATLAB-Simulink

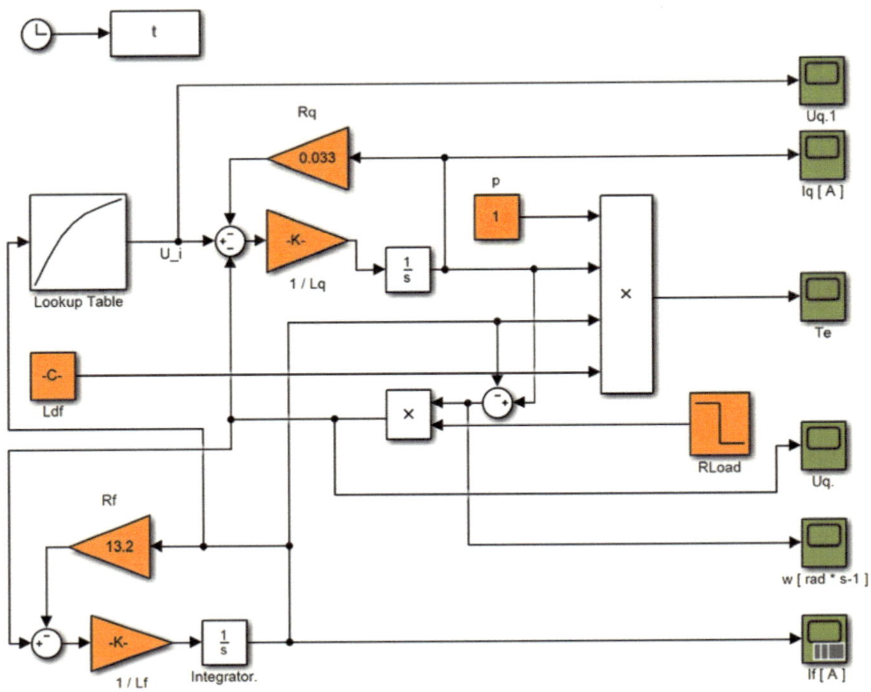

A.5 Simulation model of DC series motor in MATLAB-Simulink

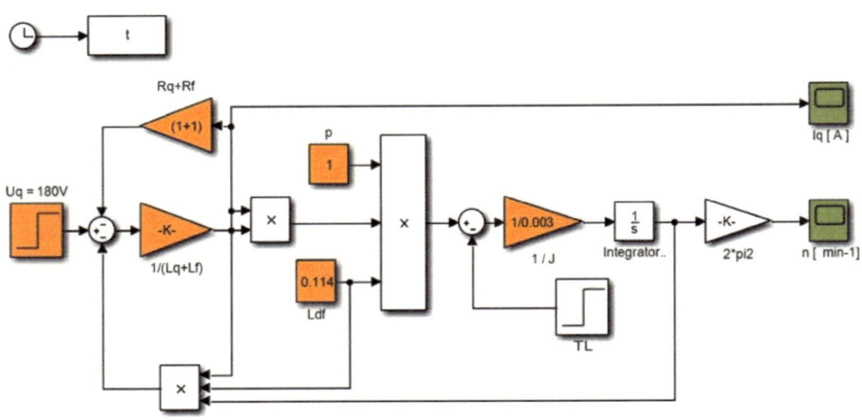

A.6 Simulation model of DC series generator in MATLAB-Simulink

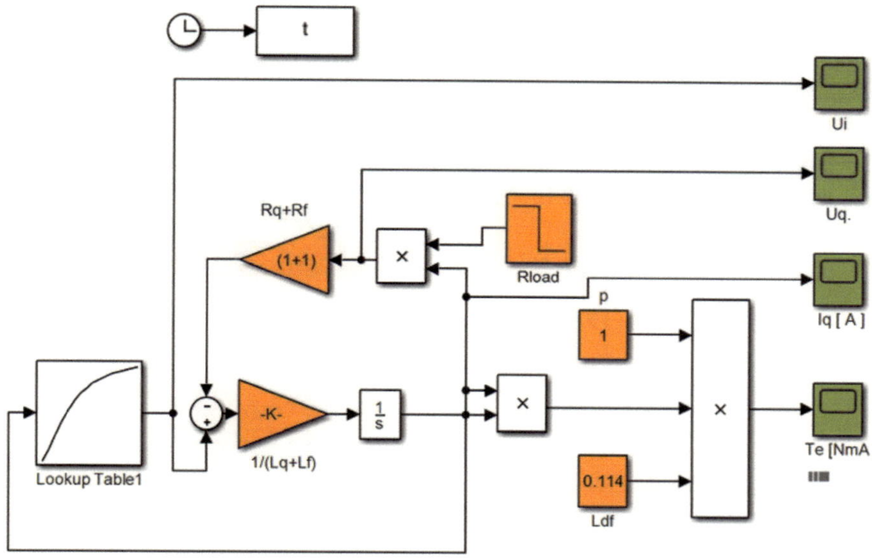

B. Appendix

B.1 Simulation model of squirrel-cage motor in MATLAB-Simulink—direct connection to the network

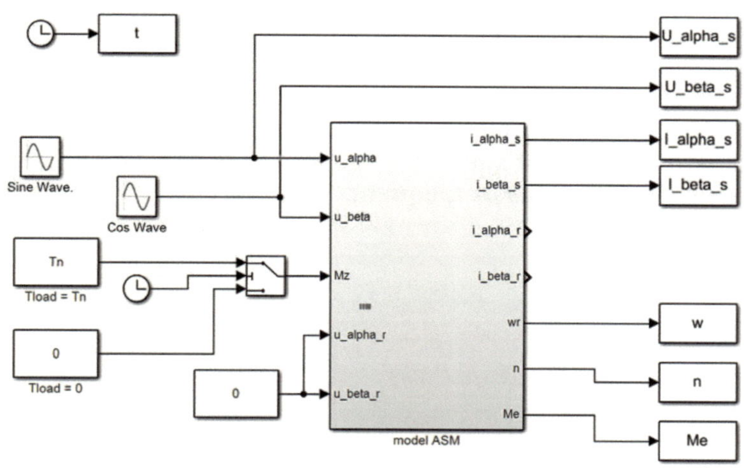

B.2 Detail of the model block of induction machine in MATLAB-Simulink

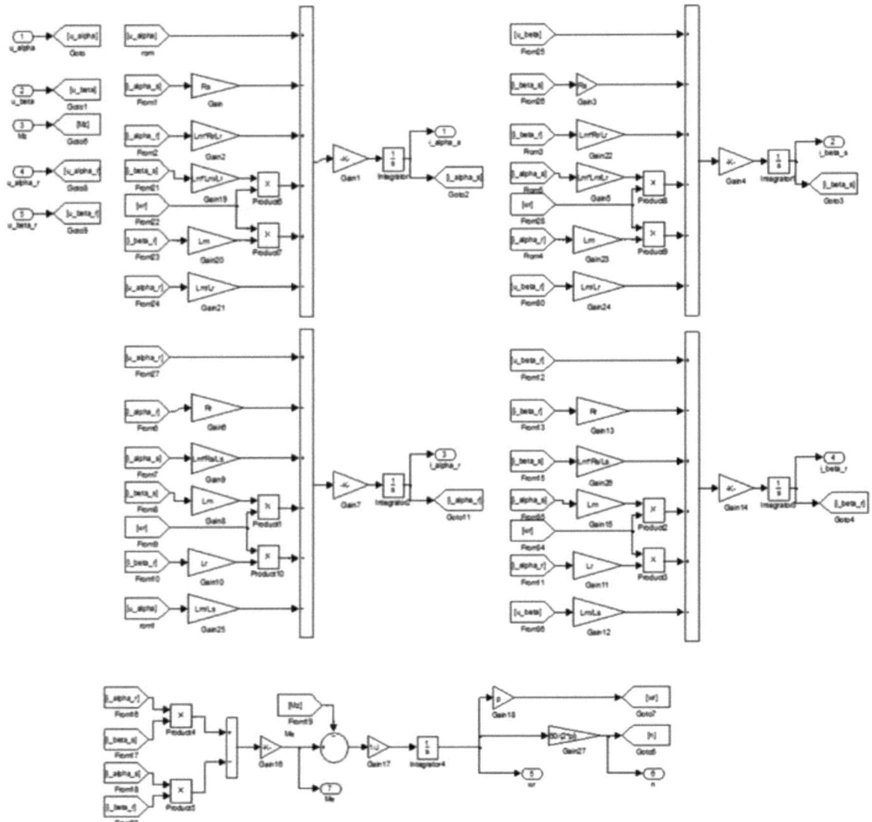

B.3 Simulation model of squirrel-cage motor in MATLAB-Simulink—starting up by means of frequency converter

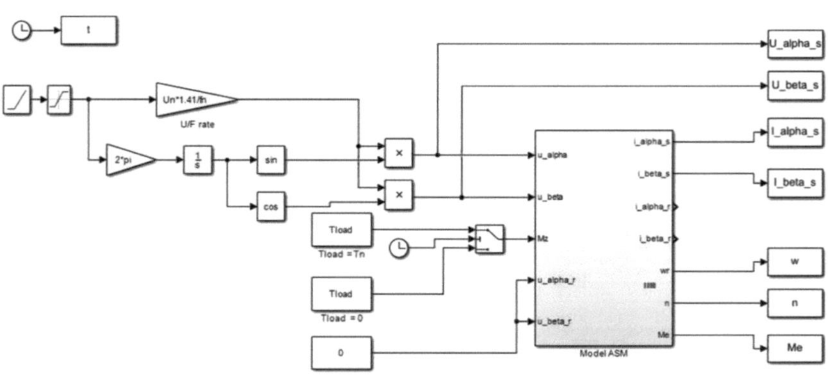

B.4 Simulation model of squirrel-cage motor in MATLAB-Simulink—starting up by means of softstarter

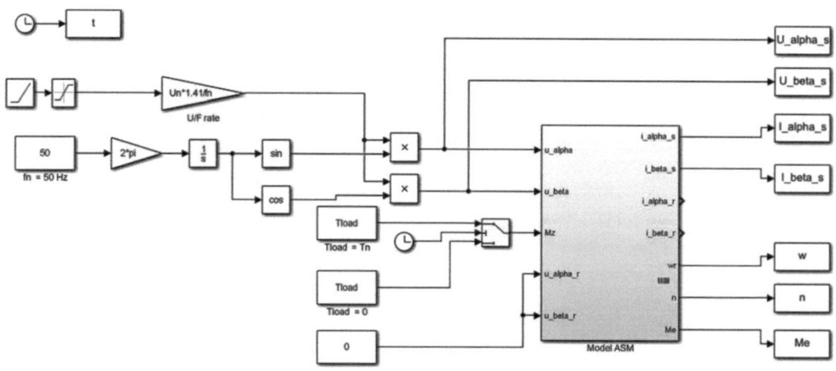

B.5 Simulation model of wound rotor asynchronous motor in MATLAB-Simulink—starting up by means of rheostat added to rotor circuit

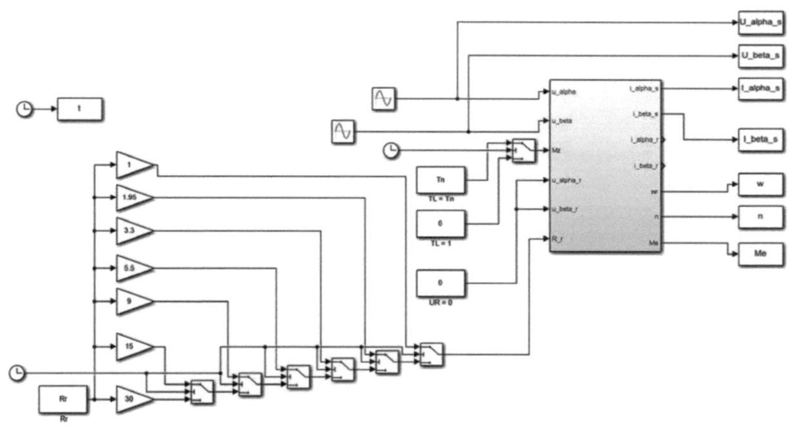

C. Appendix

C.1 Simulation model of synchronous motor with field winding in MATLAB-Simulink—starting up by means of frequency converter

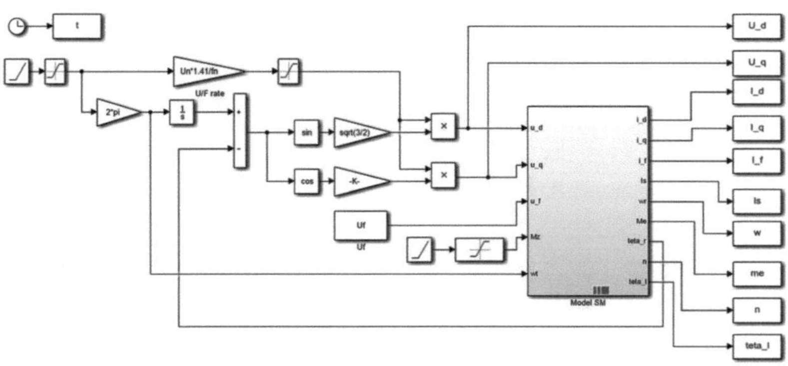

C.2 Detail of the model block of synchronous motor with field winding in MATLAB-Simulink

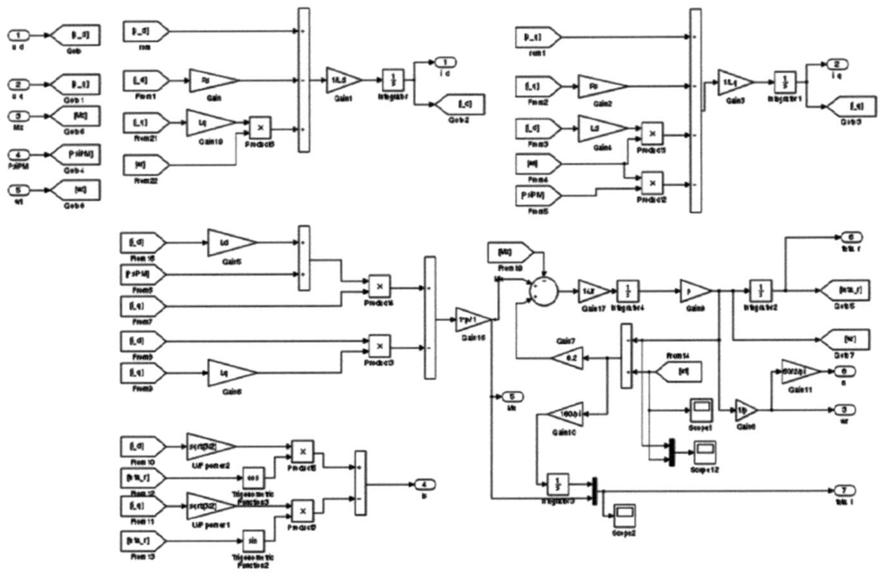

C.3 Simulation model of synchronous motor with PM in MATLAB-Simulink—starting up by means of frequency converter

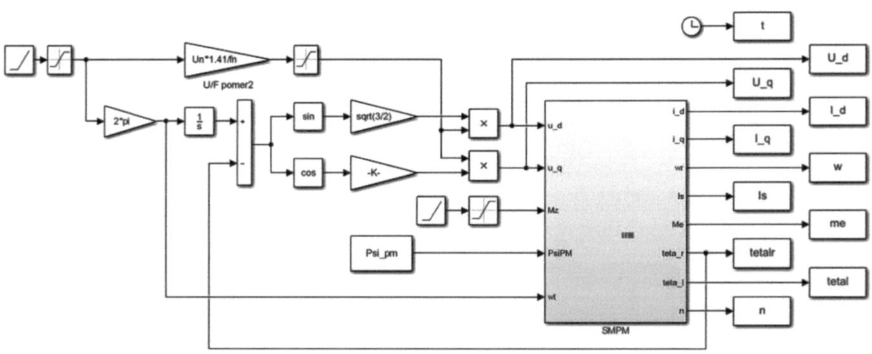

C.4 Detail of the model block of synchronous motor with PM in MATLAB-Simulink

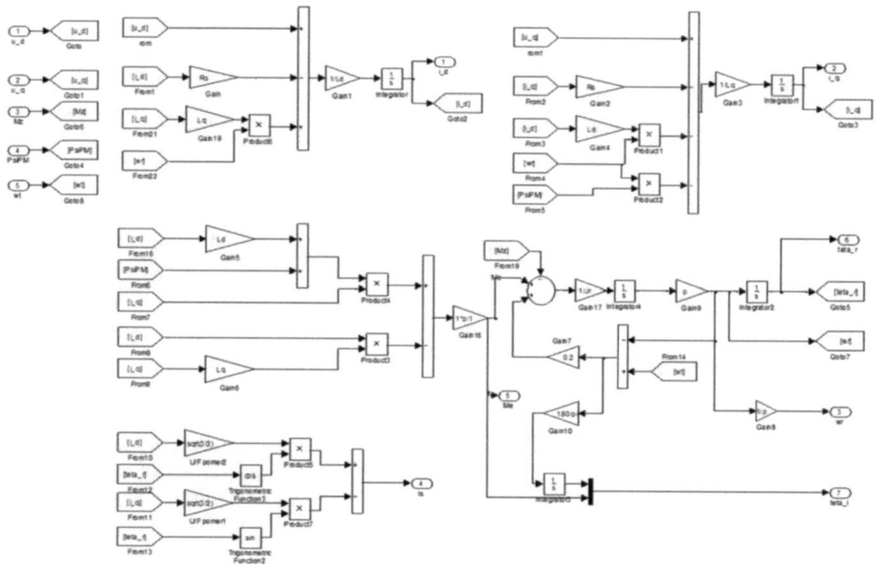

D. Appendix

D.1 Lua-Script of FEM program for computing of SRM static parameters

```
for p=1,30,1 do
        open_femm_file("c:\\Temp\\SRM\\SRM.FEM");
        modifymaterial("copper1",4,p*0.5);
        modifymaterial("copper2",4,-p*0.5);
        save_femm_file("temp.fem");
        handle=openfile("koenergia.txt","a");
        write(handle,"\n",p);
        closefile(handle);
        handle=openfile("indukcnost.txt","a");
        write(handle,"\n",p);
        closefile(handle);
        handle=openfile("temp","w");
        write(handle,p,"\n");
        closefile(handle);
        handle=openfile("moment.txt","a");
        write(handle,"\n",p);
        closefile(handle);
        handle=openfile("temp1","w");
        write(handle,0);
        closefile(handle);
          for t=0,22.5,1 do
                analyse(1);
                selectgroup(1);
                moverotate(0,0,1,(4));
                save_femm_file("temp.fem");
```

```
                    runpost("srmpost.lua","-windowhide");
                end
end
handle=openfile("temp","r");
m=read(handle,"*n");
closefile(handle);
groupselectblock(2);
l=blockintegral(0);
clearblock();
handle=openfile("indukcnost.txt","a");
write(handle," ",l*45/(m*m));
closefile(handle);
groupselectblock(3);
groupselectblock(1);
groupselectblock(2);
w=blockintegral(17);
handle=openfile("koenergia.txt","a");
write(handle," ",w*45);
closefile(handle);
handle=openfile("temp1","r");
n=read(handle,"*n");
closefile(handle);
handle=openfile("moment.txt","a");
write(handle," ",(w-n)*2578.31008);
closefile(handle);
handle=openfile("temp1","w");
write(handle,w);
closefile(handle);
exitpost()
```

Author details

Valéria Hrabovcová*, Pavol Rafajdus and Pavol Makyš
University of Žilina, Žilina, Slovakia

*Address all correspondence to: valeria.hrabovcova@feit.uniza.sk

References

[1] Pyrhonen J, Jokinen T, Hrabovcová V. Design of Rotating Electrical Machines. Chichester, West Sussex, United Kingdom: John Wiley & Sons Ltd.; 2008. ISBN: 978–0–470-69516-6, second edition 2014

[2] Vogt K. Elektrische Maschinen, Berechnung rotierender elektrischer Maschinen. Berlin: VEB Verlag Technik; 1974. in German

[3] Cigánek L. Design of Electrical Machines. In: SNTL Prague. 1958. in Czech

[4] Hindmarsch J. Electrical Machines and their Application. 4th ed. Oxford: Pergamon Press; 1991

[5] Vas P. Parameter Estimation, Condition Monitoring, and Diagnosis of Electrical Machines. Oxford: Clarendon Press; 1993

[6] Hindmarsch J. Electrical Machines and Drives, Worked Examples. nd ed. Oxford: Pergamon Press; 1985

[7] Adkins B, Harley RG. The General Theory of Alternating Current Machines, Application to Practical Problems. London: Chapman & Hall; 1975. Reprinted 1979

[8] Měřička J, Zoubek Z. General Theory of Electrical Machine. SNTL; 1973. in Czech

[9] Ong CM. Dynamic Simulation of Electric Machinery, Using Matlab/ Simulink. Purdue University, Prentice Hall; 1998

[10] Meeker D. Manual of FEMM software, web page: http://www.femm.info/Archives/doc/manual.pdf

[11] Bianchi N. Electrical Machines Analysis Using Finite Elements. Boca Raton: Taylor and Francis; 2005

[12] Szántó L. Maxwell equations and their derivation. In: Technical Literature BEN, Prague. 2003. in Czech

[13] Adkins B, Harley RG. The General Theory of Electrical Machines. London: Chapman & Hall; 1957. Reprinted 1959

[14] Park RH. Two-reaction theory of synchronous machines. Generalized method of analysis—Part I. AIEE Transaction. 1929

[15] Gieras JF, Wing M. Permanent Magnet Motor Technology, Design and Applications. 2nd ed. 2002. Revised and expanded

[16] Miller TJE. Brushless Permanent-Magnet and Reluctance Motor Drives. Oxford: Clarendon Press; 1989

[17] Sekerak P, Hrabovcova V, Pyrhonen J, Kalamen L, Rafajdus P, Onufer M. Comparison of synchronous motors with different permanent magnet and winding types. IEEE Transactions on Magnetics. 2013;**49**(3):1256-1263

[18] Hudák P, Hrabovcová V. Mathematical modelling and parameter determination of reluctance synchronous motor with squirrel cage. Journal of Electronics and Electrical Engineering. 2010;**61**(6):357-364. ISSN 1335–3632

[19] Boldea I. Reluctance Synchronous Machines and Drives. Oxford: Clarendon Press; 1996

[20] Hudák P, Hrabovcová V, Rafajdus P. Geometrical dimension influence of multi-barrier rotor on reluctance synchronous motor performances. In: International Symposium on Power Electronics, Electrical Drives, Automation and Motion, SPEEDAM 2006. 2006. pp. S42-24-S42-29

[21] Dolinar D et al. Calculation of two-axis induction motor model parameters using finite elements. IEEE Transactions on Energy Conversion. June 1997;**12**(2): 133-142